Social Media for Project Management

Edited by

Johan Ninan

Assistant Professor
Faculty of Civil Engineering and Geosciences
Delft University of Technology (TU Delft), Netherlands

CRC Press
Taylor & Francis Group
Boca Raton London New York

CRC Press is an imprint of the
Taylor & Francis Group, an **informa** business
A SCIENCE PUBLISHERS BOOK

First edition published 2022
by CRC Press
6000 Broken Sound Parkway NW, Suite 300, Boca Raton, FL 33487-2742

and by CRC Press
4 Park Square, Milton Park, Abingdon, Oxon, OX14 4RN

CRC Press is an imprint of Taylor & Francis Group, LLC

Library of Congress Cataloging-in-Publication Data (applied for)

ISBN: 9781032103785 (hbk)
ISBN: 9781032103877 (pbk)
ISBN: 9781003215080 (ebk)

DOI: 10.1201/9781003215080

Typeset in Palatino
by Radiant Productions

Preface

The number of projects is increasing worldwide as traditional and repetitive tasks are carried out through automation. Projects, being temporary and unique while being adopted globally across sectors, present a challenge for the effective management of environmental, economic, and social parameters. These temporary organizations are people-centric and require an effective management of internal and external stakeholders. In this modern age, social media is seen as a tool that connects people across the world and has significant implications on everyone's daily life. Social media is used for different purposes and encompasses multiple affordances because these are often free and bring together people from different walks of life as they tend to use them differently. However, the role of social media in managing projects is still underexplored. In this book, multiple authors, working on the application of social media in projects, come together to craft an agenda for the future. The importance of social media in projects for both internal and external stakeholders is discussed along with the avenues for research projects through social media, such as for project community engagement, for engaging internal stakeholders, for operations of projects, etc. The different social media and some of the methodologies to study them are explored in this book.

The book contains chapters written by leading scholars of the world on social media for project management. The authors are from Australia, Brazil, China, Germany, Finland, India, the Netherlands and the United Kingdom. Most of the authors have received recognition for their work on social media for project management. Dr Sunila Lobo received the 2021 Paper of the Year Award from Project Management Institute's (PMI's) *Project Management Journal (PMJ)* for her paper on social media. Dr Jere Lehtinen received the 2021 Global Young Researcher Award from the International Project Management Association (IPMA) for his research on social media for external stakeholders. I too received the 2020 Paper of the Year Award from the Association for Project Management (APM), sponsored by the *International Journal of Project Management (IJPM)*, for my paper on branding practices of projects in social media. The widespread recognition to scholars working on social media from leading project management practitioner bodies, such as PMI, IPMA, and APM, along with academic and research journals, such as *IJPM* and *PMJ*, is a testament to the importance of social media for project management in the modern digital era.

Thus, this book brings together multiple authors to discuss how social media can be used in project settings to facilitate interactions and strategic conversations across hierarchical levels and geographic boundaries for diverse goals. The book will prove to be a valuable resource for all project management academics, researchers, and practitioners who are interested in learning about the application of social media in project settings.

Contents

1

The Past, Present and Future of Social Media in Project Management

Johan Ninan

1.1 What is Social Media?

In the modern era, people spend an increasing period of time online, either communicating with others or searching for information or entertainment. As per recent estimates, around 3.6 billion people were active on social media worldwide in 2020 and this number is forecasted to reach almost 4.41 billion by 2025 (Statistica, 2022). An individual, on an average, spends more than two hours per day on social media (*Global Web Index Social Flagship Report*, 2019) as he/she increasingly uses it for connecting with his/her friends and colleagues as well as for entertainment, news, information, socializing, and self-status seeking (Park et al., 2009). Research on organizations claims that social media is also used for creating gradual political change, animating social movements, and sometimes even toppling and forming governments, as seen with the Cambridge Analytica scandal in which psychographic profiles were developed, based on users' social media data to strategically influence behavior by using targeted political advertisements (Cadwalladr and Graham-Harrison, 2018).

Information can be exchanged online in two ways: Web 1.0 considers only one-sided exchange of information, while Web 2.0 considers a two-way exchange of information. Social media can be defined as 'a group of internet-based applications that build on the ideological and technological foundations of Web 2.0, and that allow the creation and exchange of user-

Delft University of Technology (TU Delft), Netherlands.

generated content' (Kaplan and Haenlein, 2010). It involves peer-to-peer networks which are corporative, decentralized, and network pushed (Dufty, 2012). Social media covers a broad range of tools that enable the two-way exchange of information. Twitter, Instagram, and Facebook are often highlighted as the most common forms of social media. However, many one-way communication mediums are now being extended to enable two-way communications and hence can also be classified as social media. For example, digital news articles used for providing one-way information to its readers currently also enable replies and engagement from its readers and can be classified as social media (Ninan et al., 2022). Additionally, e-mails, which were considered as part of one-on-one digital communication, are currently bordering social media with the advent of group e-mails and their replies. It is also important to recognize that while social media can easily be categorized as a digital tool, it is wholly about people (Marken, 2007) as to how they use technology for establishing and maintaining relationships. So, in contrast to traditional media, social media enables multiple-way communications, compiles user-generated evolving content, allows active involvement, and is less expensive (Nikolaidou and Papaioannou, 2018).

After introducing social media, this chapter discusses the advantages of social media and why projects need social media for addressing some of its concerns. Then the chapter summarizes some current research in social media, covering internal stakeholders, external takeholders, and guidelines for education and research. This chapter also provides some directions for future research and development in the use of social media for project management, both by practitioners and researchers.

1.2 Advantages of Social Media

The advantages of social media can be classified into its reach, its different forms and tools, and its diverse impacts. Each of these is described below.

1.2.1 Reach

The biggest advantage of social media emerges from its digital transmission of data to the masses. It is a vehicle to transmit information to the public and to the traditional media in the shortest time-frame (Queensland Police Service, 2011). The subsequent reduced costs of coordination and the non-hierarchical nature of interaction are highlighted as some unique advantages of social media (Boyd and Ellison, 2010). This enables social media to act as a platform where the common man can raise his voice—be it to express himself or raise his concerns.

Social media platforms also enable the collection and analysis of real-time public discussions related to an event (Kwak et al., 2010). The discussions allow comparisons between different perspectives of an issue (Zaglia, 2013) and therefore, it is a platform that allows non-noisy stakeholders, who do not have the means to conduct or rally or express themselves in the news media, to be heard (Royer, 2000). Since the platform is part of Web 2.0, it offers one-to-one, one-to-many, or many-to-many interactions between users who can create, archive, and retrieve user-generated content (Salmons, 2014). In social media, communication is interactive, networked, and fluid between those who generate and receive the content (Bechmann and Lomborg, 2013).

The reach of social media is amplified as these platforms have an emotional echo-chamber effect, wherein emotions can be freely and publicly expressed and become amplified as others react to them, and express their own emotions (Toubiana and Zietsma, 2017). A viral social media post by a group of actors can disrupt organizations and their resource allocations, either positively or negatively. Mobile social media on smartphones further expands the reach of social media by providing a ubiquitous communication environment at the fingertips that enable people to stay connected and keep updated anytime and anywhere (Bingham and Conner, 2015). The reach of social media is significant considering the ease of use, sense of control, and the opportunity for all users to learn and improve (Ritz et al., 2019).

1.2.2 Different Forms and Tools

Different forms of data, such as text, photos, and videos are supported in social media platforms. Currently emojis, gifs, and memes are also being widely used in the social media space. This is in stark contrast with conventional media outlets, such as newspapers, which have only text and pictures. Different combinations of the data, such as video and text, picture and text, video and picture, etc., can have different implications. For example, the use of video and text in a social media page increases trustworthiness of the service rendered (Filieri, 2016).

Different social media tools can also be used for different purposes, for example, LinkedIn is about identity and reputation and is used for jobs; Facebook enables building and maintaining relationships and is used for connecting friends; YouTube enables creating and sharing video content along with user engagement through comments; Twitter enables 140-characters short messaging and is used for conversations; Quora involves questions and answers and hence is a repository of peer knowledge; Instagram enables sharing of photos and acts as a photo blog, etc. There are seven broad categories of social media, as recorded in Sterne (2010) and Nikolaidou and Papaioannou (2018). First, forums and message social media enable an exchange of thoughts, opinions or experiences. These include social media websites, such as Quora, Google groups, and the erstwhile Yahoo answers. Second, review and opinion social media enable sharing reviews, feedbacks, and opinions regarding a particular service. These include social media websites, such as TripAdvisor. Third, social networks/social media connect people who share common interests. These includes Facebook for networking among friends or LinkedIn for professional networking. Fourth, blogging social media involves posting news or sharing personal opinions and interests. These include WordPress and LiveJournal. Fifth, microblogging social media involves a shorter version of blogging, such as the 140-character short posts in Twitter. Sixth, bookmarking social media involves saving, organizing, and sharing internet bookmarks. These include Del.ici.ous and StumbleUpon. Finally, media-sharing social media involves a dedicated space for photos and videos, for example, Instagram for photos and YouTube for videos. The trust level can be different for each social media platform; for instance, consumers may trust the reviews and ratings published in Facebook more than in Tripadvisor.com, because the reviewer can be a person in their social network with a public profile, where other information as well is provided (Dwivedi et al., 2021). Thus, the choice of social media platforms depends on the use case of the consumers (Thompson, 2017).

1.2.3 Diverse Impacts

Social media is an evolving area and it can have diverse impacts and serve a variety of purposes (Lindsay, 2011). Traditionally, social media was used to connect friends, but now are used in diverse use cases, such as, for example, poling for decision making. Since transactions on social media are now computer-mediated and indexed, researchers and practitioners can now observe behaviors that were previously unobservable, and which can be modelled and managed (Varian, 2014).

Social media can be a tool to build or to destroy. It can trigger disruptions and lead to legitimacy crises for organizations, institutions, and realms, and destabilize existing arrangements (Barros, 2014). Social media has been identified to have played an important role in the Arab spring, destabilizing existing institutional arrangements in countries (Comunello and Anzera, 2012) and is a frequent tool for protests and activism (Valenzuela, 2013). With the advent of social media, powerful organizations and institutions are being brought down by people who used to be powerless, such as bloggers or even anyone with a social media account (Ronson, 2015).

Social media can also be used for marketing and branding. Organizations are no longer confined to traditional advertising, such as printed media, banners, publicity, or endorsements and now leverage the power of Web 2.0 through social media (Sivertzen et al., 2013). Social media is the place where people are connected for entertainment and a social media advertisement or engagement can have a lasting impact on the users (Ninan et al., 2019). Organizations engage with consumers and influence consumer perceptions of the brand in these social media websites or apps (Brodie et al., 2013). By using user-generated content, social media enhances the bond between the consumer and the organization (Geurin and Burch, 2017) by influencing consumer's attitudes, receiving feedbacks to improve products or services (Algharabat et al., 2018), and thereby achieve brand goals, such as improved brand awareness and increased sales at lower costs (Dwivedi et al., 2021).

Social media thus is more than just a set of tools to use; it is a larger network of people tied together through virtual and offline connections (Freberg, 2021). There can be other impacts of social media which we are still unaware of due to lack of research or the rapidly changing nature of the media. Dwivedi et al. (2021) highlight that social media is on a perpetual beta mode with the consistent release of new functions and platforms, making change a constant. The fast rate of change in social media can lead to different bloom cycles resulting in fragmented knowledge across different social media platforms (Felix et al., 2017). Even for impacts, such as the well-researched marketing impacts of social media, Canovi and Pucciarelli (2019) record that owners recognize its social, economic, and emotional benefits, but are far from exploiting its full potential.

1.3 Why Projects Need Social Media?

With the advent of robotics and automation, most of the workforce is engaged in projects resulting in projectification (Jensen et al., 2016). A project setting has people working for the project at all levels, such as managers and engineers, or those being affected by the project, such as communities and project-affected persons (PAP) (Tutt and Pink, 2019). Due to this, Packendorff (1995) calls project management as a social activity with people at the centre of all interactions. With the advent of internet, escalated by the COVID-19-induced lockdowns, many interactions between people regarding projects are in social media.

Most megaprojects have a social media presence, either by their own initiative in support of the project or by some special interest groups against the project. Ninan et al. (2019) studied social media posts on Twitter and Facebook and highlighted the effect it has on the project community. The authors highlight how posts that appeal to the community are instrumental in changing the community to become advocates of the project. Along with semi-structured interviews, the authors show how such social media interactions were instrumental in winning support for project activities on-site. Lehtinen (2021) calls projects to plan a communication strategy by framing social media discourses around issues of sustainability and entertainment, all the while using a positive, inclusive, and helpful tone. In project settings, social media posts along with its comments and likes can be a valuable source of data to track member reaction and the organization's response to it (Toubiana and Zietsma, 2017). More in-depth research, using online naturalistic data, can help us understand the digital construction of organizational identity (Johansen, 2016) in projects.

Social media can be used by stakeholders, including employees and community members, who have traditionally been marginalized to voice their concerns to organizations in a way that is visible to others (Lewis et al., 2014). Lobo and Abid (2020) studied Facebook and Twitter posts of the interest groups that act against the project and how they mobilize to achieve their vested interests. In their study, an author became a member of the interest group's Facebook site as a non-participant observer to receive alerts when new posts are added. They highlight the use of social media by the interest groups to cohere, empower, and mobilize stakeholders and thereby, create a strong collective identity against the project. Thus, social media offers plenty of online naturalistic data for researchers in the project management field to study multiple interactions, which can help the research community to understand how projects and resistance groups organize and disorganize. Also, the multiple data support in social media, which includes textual, video, interactive, affords more opportunities into gaining insights into project organizing (Ninan et al., 2020). Videos and images of implementation, including aerial views and drone footage of the construction site, can increase social media reactions, comments, and shares (Lehtinen, 2021).

There are also unintended impacts while using social media; for example, in their study of the use of social media for external stakeholder management in an infrastructure project, Ninan et al. (2021) record the impact on project personnel rationalities, such as the perception of the project as important with an overemphasis on time and an acceptability of public inconvenience. For internal stakeholders, Winter and Chaves (2017) registered an increase in the productivity of the project teams using social media. Social media can also be used for providing real-time information for project management decisionmakers. Project managers expect information in the real time and social media provides information in different forms and formats much faster, thus enabling monitoring and updating information in real-time through more intuitive interfaces (Levitt, 2011). Social media can also be used for requirements management, such as by sharing acceptance criteria and photos for approvals (Daemi et al., 2020). For internal stakeholders, social media can also be used for effective communication, knowledge sharing, and coordination (Ma et al., 2021).

Thus, social media can be used for multiple purposes in project settings including community discussions (Lobo and Abid, 2020; Williams et al., 2016), providing real-time updates on the project (Ninan et al., 2019), public opinion mining (Zhou et al., 2020), researching best practices (Ninan et al., 2020), building a targeted community for collaboration, requirements management (Daemi et al., 2021), and even for evaluating the benefits of a project (Mathur et al., 2021). Since data is considered as the new oil, social media in projects can be considered as a new ore that should be mined, processed, and used for the benefit of society. Practitioners can harness the power of society media to create projects that provide benefits for society. Researchers can use social media to study the interaction between people across project settings to understand projects and improve their practice (Ninan, 2020). More research is needed to understand the diverse impacts and use cases of social media in project settings—an envelope this book seeks to expand.

Even with the diverse use cases of social media in project settings, research by Azhar et al. (2019) highlights that social media applications are not utilized widely in the construction project sector. Some of the barriers to the use of social media in project settings include lack of corporate strategies and policies, the requirement of increased resources, a lack of project teams buy-in, the possibility of information overload, privacy considerations, and the lack of a proven business case (Harrin, 2016). However, there is increasing evidence that the pace of adoption of new ICT in construction is often determined by the benefits perceived by users and other project stakeholders (Gajendran and Perera, 2017). Hence it is important for researchers to study the diverse impacts of social media and how social media can contribute to making projects generate more value for the society.

1.4 Past, Present and Future Directions

Social media is used for understanding and addressing some of the concerns in project settings. The review of existing research and directions for future research, which constitute the contents of this book, is depicted in Fig. 1.1.

This book consists of 12 chapters with this chapter acting as the introduction and Silvius (Chapter 12) being the conclusion chapter. Firstly, chapters discussing social media use for internal stakeholders include Thompson (Chapter 2), Chaves and colleagues (Chapter 3), Jia and colleagues (Chapter 4) and Harrin (Chapter 5). Secondly, chapters discussing social media use for external stakeholders include Lobo (Chapter 6), Ninan and colleagues (Chapter 7), Zhou and colleagues (Chapter 8), and Lehtinen and Aaltonen (Chapter 9). Finally, the book

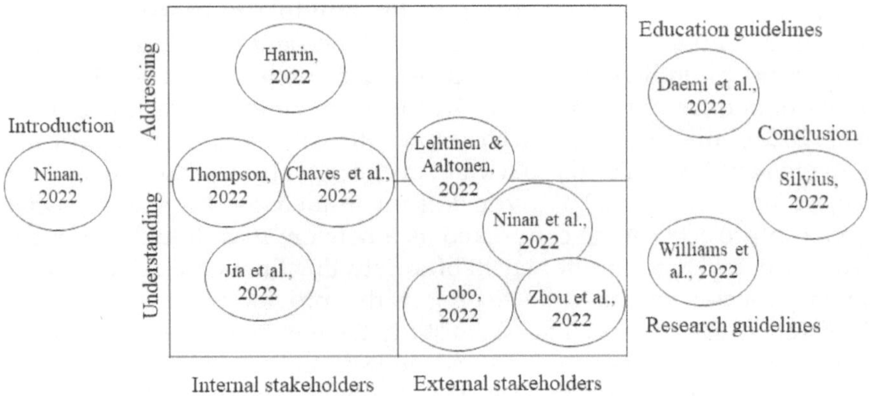

Fig. 1.1: Structure of the book.
(Alt Text: Structure of the book shown in a matrix framework covering internal stakeholders, external stakeholders, education guidelines, and research guidelines)

also has chapters that discuss guidelines for education regarding social media by Daemi and colleagues (Chapter 10), and guidelines for research on social media by Williams and colleagues (Chapter 11). Within all these three domains, there is a need to understand practices of the project and behaviors of the stakeholders to address concerns and improve projects.

We also cover the use of social media by different projects. Chaves and colleagues (Chapter 3) study social media use by Information Technology projects and Lobo (Chapter 6), Zhou and colleagues (Chapter 8), and Ninan and colleagues (Chapter 7) study social media use by infrastructure megaprojects. Different social media are also covered in this book. Ninan and colleagues (Chapter 7), Lobo (Chapter 6) studies Twitter, Harrin (Chapter 5) studies Facebook and Clubhouse, and Zhou and colleagues (Chapter 8) study SINA Weibo. While Lobo (Chapter 6) and Ninan and colleagues (Chapter 7) use qualitative analysis of social media data, Zhou and colleagues (Chapter 8) use quantitative data analysis of social media data.

1.4.1 Internal Stakeholders

Internal stakeholders, such as the organization's employees, the contractor building the project, or the designer of the project, have a contractual relationship with the owner of the project and reply to it to enforce a claim (Winch et al., 2007). Social media can help to understand the internal stakeholder issues and address them.

Thompson (Chapter 2) discusses how digitalization in projects through social media can improve project management practice. She uses theories from systems thinking and organizational learning to understand the practice of communication among internal stakeholders in project settings. The types of social media used and the different activities in a project they are used for are uncovered. She then proposes that social media can be used to address some of the current issues in internal stakeholder management, such as digital sharing of tacit and factual knowledge and thereby learning in project settings.

Chaves and colleagues (Chapter 3) discuss how social media is used in the home/office work period due to COVID-19 from a senior project worker's perspective. They distinguish between active, passive, formal, and informal communications and map which social media tool is used for each of these in project settings. They highlight that social media can address some of the challenges by providing a transparent medium for communications and knowledge sharing. They record the presence of a team culture in the use of social media for internal stakeholders and call for more research on improving information security, confidentiality, and developing a formal knowledge-management process.

Jia and colleagues (Chapter 4) discuss how social media is used for knowledge management considering the temporary and urgent nature of project settings. Social media can enable knowledge management at the individual, team and project levels by leveraging its visibility, editability, persistence, and association. They highlight how social media is used for knowledge acquisition, knowledge sharing, and knowledge integration.

Harrin (Chapter 5) considers how social media can be used for addressing some of the challenges of internal stakeholders in project settings. The chapter offers a reflexive account of her role in the informal mentoring of project employees in multiple curated online spaces, such as Facebook and Clubhouse. She notes that social media offers informal support, peer-to-peer mentoring, continuous professional development, and knowledge-sharing opportunities for project delivery professionals. In the chapter, she offers some guidelines and practical tips on setting up and running a successful project management professional's community in social media.

There is more work to be done on understanding and addressing the concerns of internal stakeholders in project settings. Social media and its multiple affordances (Ninan et al., 2020) offer significant opportunities to transform the project environment, both in improving its productivity and work culture. Future research can study social media platforms, such as LinkedIn and project management specific groups to understand recruitment, the impact of project management certifications, and the role of peer testimonials in influencing recruitment.

1.4.2 External Stakeholders

External stakeholders include stakeholder's peripheral to the project, such as owners of the land from whom the land is acquired, those who are inconvenienced by the construction noises, vibrations, diversions, etc. and those who stand to benefit from the project in improving the services (Ninan et al., 2021; Viitanen et al., 2010). Since they do not have any contractual relationship with the client, they rely on regulators, political influence or public campaigns to enforce a claim (Winch et al., 2007). Social media can help to understand the external stakeholder issues and address them.

Lobo (Chapter 6) investigates the role of social media in managing external stakeholders in the High Speed Two (HS2) project, UK. The Twitter posts of wildlife and environmental activists are studied to understand the concerns of these external stakeholders. She notes that the issues that concern external stakeholders can be discerned by examining posts on social media which could be used to create a framework for project managers to better manage interactions with and responses to external stakeholders.

Ninan and colleagues (Chapter 7) highlight how social media can be used to improve project operations considering the case of metro rail projects in Australia and India. They develop a framework for engaging with the users of an infrastructure project during the operation phase, using the big data from social media. The Twitter data can be used to identify day-to-day technical issues, collate suggestions to improve, and monitor live sentiments of the project. They even propose a data-science initiative-delivery framework for capturing the voice of customer during the operation phase.

Zhou and colleagues (Chapter 8) record how online public opinion can be understood in megaprojects considering the Hong Kong-Zhuhai-Macao Bridge. Web crawler was adopted to acquire data from SINA Weibo social media platform to understand the sentimental tendency and topic variation across different stages of the project. They note that public opinion analysis during the earlier stages of a project can provide reference to public opinion guidance and prediction in the future, thereby improving the performance of the megaproject.

Lehtinen and Aaltonen (Chapter 9) discuss how social media can be used for value creation and distribution, considering the external stakeholders. They develop a conceptual framework of engagement and communication that project organizations can adopt when using social media for external stakeholder engagement. The DICE model (Disengager, Informer, Communicator, Enfranchiser) captures the different project organization's roles in using social media for external stakeholder engagement.

More research is required to address the concerns of external stakeholders in project management. Social media offers a unique platform as it provides the necessary reach and flexibility required to bring together the vast and diverse external stakeholders together to talk about the megaprojects. Future research can explore how external stakeholders engage in these social media pages and how more value can be created for the society by projects playing an active role in these social media engagements.

1.4.3 Guidelines for Education and Research

In the modern era, people spend a lot of time on social media platforms, both for entertainment and work. As discussed earlier, social media is the future of project management since projects are interventions in society with multiple interactions with people. There is a need for guidelines to educate future managers on the effective use of social media. Additionally, the rapid growth in digital data enabled by social media provides ample research opportunities (Wiedemann, 2013). There is a need for guidelines for researchers in using social media.

Daemi and colleagues (Chapter 10) mention lack of experience and confidence by project professionals in using social media to address which they suggest a social media unit for project management students. A well-defined unit to teach social media use in project management is recommended to build the necessary social media competency. They propose various topics that can form part of the social media for project management unit. There need to be more concrete guidelines on social media education by institutions and social media data use by researchers since their use in different forms is increasing across projects. Project management educators have to include social media as part of the curriculum for professionals in the area. The syllabus should be open-ended, considering the multiple affordances and diverse use of social media platforms. Education should aim to cultivate an adaptive mindset, a willingness to engage in continuous learning, and an ability to visualize and implement social media initiatives, among the project practitioners. As researchers and educators in the area of project management, it is our duty to support practitioners with insights and to include topics relevant to the modern era in our curricula.

Guidelines for practice is another area that needs to be developed for social media. Practices of manipulating social media data through fake news and reviews for an organization's vested interests is prevalent. For example, a hotel was caught posting several positive reviews for its hotel by acting as a customer and was then ordered to pay $3 m for manipulating TripAdvisor reviews (*The Guardian*, 2018). Suitable guidelines for practices have to be set up for the project's use of social media. There is an urgent need for integration of ethical and responsible use of social media to achieve the strategic objectives of the organization. It should place people at its centre and meet their expectations.

Williams and colleagues (Chapter 11) highlight the need for researchers to also be responsible while using social media data. While there is a focus on reproducibility and inspection of the data for improving the credibility of findings, this can have moral and ethical implications. They develop a framework for responsible use of this data to support research and practice in project management. More researchers are being drawn to social media data, considering its rich, longitudinal, and digitally indexed nature (Ninan, 2020). There is a range of traditional and modern qualitative and quantitative approaches, such as questionnaires, interviews, focus groups, observation, secondary analysis, official statistics, and ethnography as well as more social media-specific approaches, such as content analysis applied to social media platforms, social media-based focus groups, social media netnography, and virtual ethnography, and social media analytics that can shed light on the practice of the project and behavior of stakeholders. Online naturalistic inquiry or ethnographic practices on social media can inform project management practices in cases where informants are not all that helpful, either by being not willing to share information or when they don't know the information (Gehman et al., 2018). Advances in digital research methods have to be developed to analyze the qualitative and quantitative data in social media to make sense of project management in this digitalized 21st century. The variety of research approaches to study social media is still limited and we need novel approaches to project management research (Drouin et al., 2013) to study the reach, means, and impact of social media in project settings. However, researchers have to consider the ethical and moral responsibilities of using the data for qualitative and quantitative analyses.

Finally, education, practice, and research should take a transdisciplinary perspective, equipping every person with the ability to develop the skills to successfully navigate the digital social world while considering the numerous known and unknown implications of any intervention.

1.5 Conclusion

With the advent of COVID-19 and the subsequent work from home, project businesses happen on social media and organizations can therefore no longer ignore these artifacts and needs to access digital communication as a critical data source for understanding contemporary organizational phenomena. Social media is changing the way projects are managed and researchers and practitioners need to look at social media to understand projects in the 21st century.

After defining social media, this chapter records the advantages of social media. It has extensive reach, has different forms and tools, and has diverse impacts. Due to these advantages, social media is used in different ways in project settings, such as progress updates, branding, and as a tool to build or destroy a project. The use cases of social media for internal stakeholders and external stakeholders are introduced and subsequently detailed in the remaining chapters of this book. The need for guidelines on the use of social media for research and in education is also introduced. Within all these three domains, there is a need to understand practices of the project and behaviors of the stakeholders to address concerns and improve projects.

References

Algharabat, R., Rana, N.P., Dwivedi, Y.K., Alalwan, A.A., and Qasem, Z. (2018). The effect of telepresence, social presence, and involvement on consumer brand engagement: An empirical study of non-profit organizations, *Journal of Retailing and Consumer Services*, **40**: 139–149.

Azhar, S., Riaz, Z., and Robinson, D. (2019). Integration of social media in day-to-day operations of construction firms, *Journal of Management in Engineering*, **35**(1).

Barros, M. (2014). Tools of legitimacy: The case of the petrobras corporate blog, *Organization Studies*, **35**(8): 1211–1230.

Bechmann, A., and Lomborg, S. (2013). Mapping Actor Roles in Social Media: Different Perspectives on Value Creation in Theories of User Participation, 15, *New Media & Society*, pp. 765–781: 5.

Bingham, T., and Conner, M. (2015). The New Social Learning: Connect. Collaborate, *Work*, 2nd edition.

Boyd, D.M., and Ellison, N.B. (2010). Social network sites: Definition, history, and scholarship, *IEEE Eng. Manag. Rev.*, **38**(3): 16–31.

Brodie, R.J., Ilic, A., Juric, B., and Hollebeek, L. (2013). Consumer engagement in a virtual brand community: An exploratory analysis, *Journal of Business Research*, **66**(1): 105–114.

Cadwalladr, C., and Graham-Harrison, E. (2018). Revealed: 50 million Facebook profiles harvested for Cambridge Analytica in major data breach, March 17, *The Guardian*. https://www.theguardian.com/news/2018/mar/17/cambridge-analytica-facebook-influence-us-election [accessed on 10 January 2022].

Canovi, M., and Pucciarelli, F. (2019). Social media marketing in wine tourism: Winery owners' perceptions, *Journal of Travel & Tourism Marketing*, **36**(6): 653–664.

Comunello, F., and Anzera, G. (2012). Will the revolution be tweeted? A conceptual framework for understanding social media and the Arab Spring, *Islam and Christian-Muslim Relations*, **23**(4): 453–470.

Daemi, A., Chugh, R., and Kanagarajoo, M.V. (2020). Social media in project management: A systematic narrative literature review, *International Journal of Information Systems and Project Management*, **8**(4): 5–21.

Daemi, A., Chugh, R., and Kanagarajoo, M.V. (2021). Social media in project management: A systematic narrative literature review, *International Journal of Information Systems and Project Management*, **8**(4): 5–21.

Dufty, N. (2012). Using social media to build community disaster resilience, *Australian Journal of Emergency Management*, **27**(1): 40–45.

Drouin, N., Muller, R., and Sankaran, S. (2013). *Novel Approaches to Organizational Project Management Research: Translational and Transformational*, Copenhagen Business School Press.

Dwivedi, Y.K., Ismagilova, E., Hughes, D.L., Carlson, J., Filieri, R., Jacobson, J., Jain, V., Karjaluoto, H., Kefi, H., Krishen, A.S., Kumar, V., Rahman, M.M., Raman, R., Rauschnabel, P.A.,

Rowley, J., Salo, J., Tran, G.A., and Wang, Y. (2021). Setting the future of digital and social media marketing research: Perspectives and research propositions, *International Journal of Information Management*, **59**: 102168.

Felix, R., Rauschnabel, P.A., and Hinsch, C. (2017). Elements of strategic social media marketing: A holistic framework, *Journal of Business Research*, **70**: 118–126.

Filieri, R. (2016). What makes an online consumer review trustworthy? *Annals of Tourism Research*, 58: 46–64.

Freberg, K. (2021). *Social Media for Strategic Communication: Creative Strategies and Research-based Applications*, Sage Publications.

Gajendran, T. and Perera, S. (2017). The Australian construction e-business review. *Proceedings of CIB TG83: e-Business in Construction*, University of Newcastle, Western Sydney University, and the Australian Institute of Building, New Castle, ISBN 978-0-6480003-0-3.

Gehman, J., Glaser, V.L., Eisenhardt, K.M., Gioia, D., Langley, A., and Corley, K.G. (2018). Finding theory–method fit: A comparison of three qualitative approaches to theory building, *Journal of Management Inquiry*, **27**(3): 284–300.

Geurin, A.N., and Burch, L.M. (2017). User-generated branding via social media: An examination of six running brands, *Sport Management Review*, **20**(3): 273–284.

Global Web Index Social Flagship Report (2019). Retrieved from https://www.globalwebindex.com/reports/social.

Harrin, E. (2016). Barriers to social media adoption on projects. In: Silvius, A.J.G (Ed.). *Strategic Integration of Social Media into Project Management Practice*, pp. 106–124, IGI Global.

Jensen, A., Thuesen, C., and Geraldi, J. (2016). The projectification of everything: Projects as a human condition, *Project Management Journal*, **47**(3): 21-34.

Johansen, T.S. (2016). Countering the 'natural' organizational self on social media. pp. 72–90. *In:* Frandesn, S., Kuhn, T., and Lundholt, M.W. (Eds.). *Counter-narratives and Organization*, Routledge.

Kaplan, A.M., and Haenlein, M. (2010). Users of the world, unite! The challenges and opportunities of Social Media, *Business Horizon*, **53**(1): 59–68.

Kwak, H., Lee, C., Park, H., and Moon, S. (2010). What Is Twitter, a Social Network or a News Media? Paper Presented at the *Proceedings of the 19th International Conference on World Wide Web*, New York, NY, USA.

Lehtinen, J. (2021). External Stakeholder Engagement in Complex Projects, Aalto University.

Levitt, R. (2011). Towards project management 2.0, *Engineering Project Organization Journal*, **1**(3): 197–210.

Lewis, K., Gray, K., and Meierhenrich, J. (2014). The structure of online activism, *Sociological Science*, **1:** 1–9.

Lindsay, B. (2011). Social media and disasters: Current uses, future options, and policy considerations, *US Congressional Research Service Report*.

Lobo, S., and Abid, A.F. (2020). The role of social media in intrastakeholder strategies to influence decision making in a UK infrastructure megaproject: Crossrail 2, *Project Management Journal*, 8756972819864456.

Ma, G., Jia, J., Ding, J., Wu, M., and Wang, D. (2021). Examining the impact of social media use on project management performance: Evidence from construction projects in China, *Journal of Construction Engineering and Management*, **147**(3): 04021004.

Marken, G.A. (2007). Social media . . . The hunted can become the hunter, *Public Relations Quarterly*, **52**(4): 9–12.

Mathur, S., Ninan, J., Vuorinen, L., Ke, Y., and Sankaran, S. (2021). An exploratory study of the use of social media to assess benefits realization in transport infrastructure projects, *Project Leadership and Society*, 2: 100010.

Nikolaidou, A., and Papaioannou, P. (2018). Utilizing social media in transport planning and public transit quality: Survey of literature, *Journal of Transportation Engineering, Part A: Systems*, **144**(4): 04018007.

Ninan, J., Clegg, S., and Mahalingam, A. (2019). Branding and governmentality for infrastructure projects, *International Journal of Project Management*, **37**(1): 59–72.

Ninan, J. (2020). Online naturalistic inquiry in project management research: directions for research, *Project Leadership and Society*, 1: 100002.

Ninan, J., Mahalingam, A., Clegg, S., and Sankaran, S. (2020). ICT for external stakeholder management: sociomateriality from a power perspective, *Construction Management and Economics*, **38**(9): 840–855.

Ninan, J., Mahalingam, A., and Clegg, S. (2021). Asset creation team rationalities and strategic discourses: Evidence from India, *Infrastructure Asset Management*, **8**(2): 1–10.

Ninan, J., Mahalingam, A., and Clegg, S. (2022). Power in news media: Framing strategies and effects in infrastructure projects, *International Journal of Project Management*, **40**(1): 28–39.

Packendorff, J. (1995). Inquiring into the temporary organization: New directions for project management research, *Scandinavian Journal of Management*, **11**(4): 319–333.

Park, N., Kee, K.F., and Valenzuela, S. (2009). Being immersed in social networking environment: Facebook groups, uses and gratifications, and social outcomes, *Cyberpsychology & Behavior*, **12**(6): 729–733.

Queensland Police Service. (2011). *Disaster Management and Social Media – A Case Study*, Media and Public Affairs Branch, Queensland Police Service.

Ritz, W., Wolf, M., and McQuitty, S. (2019). Digital marketing adoption and success for small businesses, *Journal of Research in Interactive Marketing*, **13**(2): 179–203.

Ronson, J. (2015). *So You've been Publicly Shamed*, Oxford: McMillian Publishers.

Royer, P.S. (2000). Risk management: The undiscovered dimension of project management, *Project Management Journal*, **31**(1): 6–13.

Salmons, J. (2014). Online research ethics: Questions researchers ask, answers guidelines provide. *In*: Woodfield, K. (Ed.). *Social Media in Social Research: Blogs on Blurring the Boundaries*, NatCen Social Research, London.

Sivertzen, A.M., Nilsen, E.R., and Olafsen, A.H. (2013). Employer branding: Employer attractiveness and the use of social media, *Journal of Product & Brand Management*, **22**(7): 473–483.

Statistica (2022). Number of social network users worldwide from 2017 to 2025. Available at: https://www.statista.com/statistics/278414/number-of-worldwide-social-network-users/ [accessed on 10 January 2022].

Sterne, J. (2010). *Social Media Metrics: How to Measure and Optimize Your Marketing Investment*, Wiley, Chichester, U.K.

The Guardian (2018). Meriton fined $3 m for manipulating TripAdvisor hotel reviews. https://www.theguardian.com/travel/2018/jul/31/meriton-fined-3m-for-manipulating-tripadvisor-hotel-reviews#:~:text=Australian%20property%20giant%20Meriton%20has%20been%20ordered%20to%20pay%20%243,in%20breach%20of%20consumer%20law [accessed n 10 January 2022].

Thompson, K.M. (2017). *A Framework for using Social Media in the Practice of Project Management*. PhD Thesis. UK: Bournemouth University.

Toubiana, M., and Zietsma, C. (2017). The message is on the wall? Emotions, social media and the dynamics of institutional complexity, *Academy of Management Journal*, **60**(3): 922–953.

Tutt, D., and Pink, S. (2019). Refiguring global construction challenges through ethnography, *Construction Management Economics*, **37**(9): 475–480.

Valenzuela, S. (2013). Unpacking the use of social media for protest behavior: The roles of information, opinion expression, and activism, *American Behavioral Scientist*, **57**(7): 920–942.

Varian, H. (2014). Beyond big data, *Business Economics*, **49**(1): 27–31.
Wiedemann, G. (2013). Opening up to big data: computer-assisted analysis of textual data in social sciences, *Historical Social Research*, **38**(4): 332–358.
Williams, N.L., Ferdinand, N., and Pasian, B. (2015). Online stakeholder interactions in the early stage of a megaproject, *Project Management Journal*, **46**(6): 92–110.
Winch, G.M., Morris, P., and Pinto, J. (2007). Managing project stakeholders. pp. 271–289 *In:* Morris, P. and Pinto, J. (Eds.). *The Wiley Guide to Project, Program, and Portfolio Management*. New Jersey: John Wiley & Sons.
Winter, R., and Chaves, M.S. (2017). Innovation in the management of lessons learned in an IT project with the adoption of social media, *International Journal of Innovation: IJI Journal*, **5**(2): 156–170.
Zaglia, M.E. (2013). Brand communities embedded in social networks, *Journal of Business Research*, **66**(2): 216–223.
Zhou, Z., Zhou, X., and Qian, L. (2021). Online Public Opinion Analysis on Infrastructure Megaprojects: Toward an Analytical Framework, *Journal of Management in Engineering*, **37**(1): 04020105.

2

Digital Disruption

How the use of Social Media Can Improve Project Management

Karen Margaret Thompson

2.1 Introduction

Views among the project management community on the impact of social media on the practice of managing projects are diverse. In 2014, the UK's Association for Project Management (2014) asked if social media are 'a waste of time' while others see social media and project management as a 'perfect match' (Harrin, 2010). My awareness of the potential for social media to enhance project management came while I was supervising a team of young people undertaking a project and the team invited me to join their social media community as a way of receiving project updates. These early insights were exciting because I could see that the way the team was using social media was transforming communication among the team members and with external stakeholders on the project. My curiosity was piqued. I wondered whether what I observed was just a function of excellent communication by one high-performing team, or whether social media could offer something new for managing projects! The insights I gained from this early use of social media inspired the research presented in this chapter.

Before relevant literature is reviewed and the research explained, this introduction presents a new perspective on project management as

Bournemouth University Business School, United Kingdom.

a socially constructed phenomenon and a new framework for project communication is developed that enables social media to be understood in the context of projects and project management as complex systems of human interaction.

2.1.1 Socially Constructed Project Management

General Systems Theory (Boulding, 1956) was used as a starting point for understanding projects as complex social organizations. Theoretical systems and constructs are arranged in a hierarchy of complexity where communication and information processes are essential to development of an organization, both in the biological and social world. Nine levels of complexity are defined. An important characteristic of each level is the relationship between an individual system and its external environment. At level one there are basic frameworks and static structures; level two is simple dynamic systems, such as clockworks; level three is control systems, such as thermostats, and so on up to level nine that represents transcendental, unknowable systems. Human beings are classified as level seven and social organizations, such as projects, are level eight. Traditional, rational approaches to project management tend to assume projects behave as simple, static, level three control systems. Therefore, to understand use of social media in complex social organizations requires new understandings and new definitions of project and project management, reflecting the complexity of human interactions.

The definition of a social organization as 'a set of roles tied together with channels of communication' (Boulding, 1956) was used to replace the term 'organization' in Turner's (2014) definition of a project, to create a new definition of a project as 'a temporary set of roles tied together by channels of communication to achieve purposeful change' (Thompson, 2017:4). A new definition of project management was built using notions of 'art and science' and 'converting vision into reality' from Turner (1999) to give 'project management practice is the art and science of achieving purposeful change by enabling communication, coordination and integration among temporary roles to convert vision into reality' (Thompson, 2017:5). These definitions provide a new perspective on the practice of project management that emphasizes human interaction and communication in a way that can extend a conventional deterministic and purely rationale approach and enable the role of social media to be studied in complex, socially-constructed context.

2.1.2 A New Framework for Project Communication

The importance of effective communication in project management is well-recognized, but practices tend to be deterministic, focus on ... internal requirements and fail to reflect the complexity of human interaction. Communication between a system and its external environment is emphasized by understandings of projects as open systems. As external changes occur, a project needs to seek out and interpret new knowledge in order to adapt and respond. Organizations can be passive or active in the way they obtain knowledge about the external environment (Daft and Weick, 1984). Passive organizations accept whatever information is provided and interpret the external world within accepted limits that are, in the case of projects, identified as part of project definition. In contrast, organizations that actively search for data and perform trials to learn what is feasible, develop interpretations that are different from passive organizations (Daft and Weick, 1984). On projects, lack of communication beyond the project boundary has been found in research (Müller, 2003; Partington, 1997). New requirements for communication are suggested by viewing a project as a learning organization. A learning organization connects with its environment through three processes or stages: scanning (data collection), interpretation (data given meaning), and learning (action taken) (Daft and Weick, 1984). These connecting activities are generally not well represented in prescribed project management practices.

Three types of social conversations have been identified for projects: social around the project, social about the project and social within the project (Taylor, 2016). Combining the stages of organizational learning with the three types of social communication suggests a new framework with three 'zones' for communication activities: scanning and connecting the project to its environment; engagement and interpretation; and action and learning. Feedback is required between the stages of organizational learning and therefore the 'boundary' between zones needs to be permeable, as represented in Fig. 2.1 by broken lines. Social media are added to the zones of communication, the factors influencing use and the expected impacts were identified from the fields of technology adoption and knowledge management to create a framework for the research presented in this chapter, as shown in Fig. 2.2.

2.2 Literature Review

Technology adoption is a mature field of research. Models that recognize attitude influences behavior, and attitude incorporates cognitive, affective, and behavioral components (McGuire, 1969; Ajzen, 1991) are consistent with a socially constructed view of reality. Technology adoption and use in organizational settings, where the form of technology and use are specified and managed by the organization, have been widely researched and provide the outline framework as shown in Fig. 2.1. Social media blurs the boundary between personal and organizational setting for

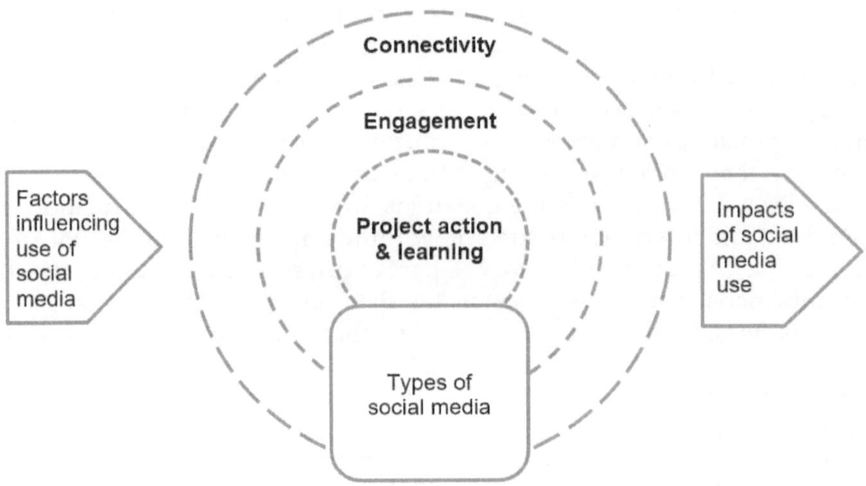

Fig. 2.1: Framework for exploring social media adoption and use on projects. (Alt Text: Framework for exploring social media adoption and use on projects, showing influencing factors, three types of communication activities, and impacts)

technology use. Although digital convergence has been widely discussed (e.g., Grudin, 2010; Hollocks, 1994), the impact of mobile technologies and social media in enabling opportunities for entirely new ways of working, enabling people to think, act, and communicate in different ways (Van der Merwe, 2016) have yet to be fully understood.

Social media is an umbrella term. For this work, social media are defined as 'the internet-based socio-technical systems, websites or applications that provide space for social interaction, communication, collaboration, and community formation' (adapted from Zhao et al., 2013).

2.2.1 Factors Influencing Technology Adoption

Research findings on the antecedents of technology use have changed over time, as technology and its uses have changed. The Technology Acceptance Model (Davis, 1989) shows that perceptions of the technology, as well as external variables, influence attitude towards using technology, and that attitude influences use through intention to use. The Unified Theory of Acceptance and Use of Technology (Venkatesh et al., 2003) includes determinants and moderating factors, classified into four groups: perceptions of the technology (effort expectancy and performance expectancy); characteristics of the individual (gender, age, experience, and voluntariness of use); social influence; and facilitating conditions. Research on collaboration technologies added task characteristics to earlier models (Bok et al., 2012; Brown et al., 2010; Goodhue and Thompson, 1995). In a project setting, Müller (2003) identifies project risk as a determinant of communication preferences. Three categories of factors are identified as influencing the use of enterprise social software platforms—technological, social, and organizational characteristics (Kügler et al., 2013). Accordingly, four categories of factors are indicated for use of social media in project settings: organizational characteristics, technological characteristics, individual and group characteristics, task, and project characteristics.

2.2.2 Social Media Use on Projects

Use of social media to enable remote working has become commonplace in recent times. Harrin (2010) suggests a range of project management activities that could be supported by use of social media, including knowledge management, collaboration and team working, meetings, project log and status updates that are needed in conventional project management. Use of instant messaging to provide fast and direct answers to questions is highlighted by Harrin (2010) and is one way that social media can support human interaction in projects. The role of instant messaging in meetings was the focus of research by Dennis et al. (2010) who found new interactions, called 'backstage conversations' and 'invisible whispering', that may have an influence on collaborative decision-making. Four phases of virtual group decision-making were used by Turban et al. (2011)—intelligence, design, choice, and implementation—and they suggest that functions of technology, such as polling, voting, and discussion groups are important for enabling interactions. Gimpel et al. (2014) research on users' perceptions of the relevance and benefits of social collaboration technologies in project work identified a range of activities. Some of the activities found by Gimpel et al. (2014) relate to project work and are similar to the suggestions of Harrin (2010): storing information and spreading content (knowledge management), coordinating project work (collaboration/team working, project log and status updates). In contrast, other activities found by Gimpel et al. (2014), such as solving problems, brainstorming, informal exchange, discussion, and organizing meetings, emphasize human interaction in ways that are consistent with a socially constructed view of projects.

Suggestions for uses of social media in project settings are also made by the APM (2014). Four main uses are identified: discursive, networking, events calendar/task scheduling, and collaboration. Three discursive activities are sharing lessons learnt, sharing best practice, and stakeholder management. Networking, using external platforms, is highlighted and they suggest social platforms can be used to source skills and expertise externally. Collaborative calendars (e.g., Doodle) and software to manage workstreams (e.g., Trello) are suggested. A role for collaboration tools and shared workspace is also suggested by APM (2014) for controlling project documentation and co-authoring.

A range of mobile apps for project management were analyzed by Silvius and Silvius (2016). They found two main application areas: (1) support for the individual project manager in the planning/organizing processes of the project and (2) support for team communication and team collaboration and Silvius and Silvius (2016) highlight the lack of functionality for communication/collaboration with project sponsor and other stakeholders. Similarly, Van der Merwe (2016) differentiates use of social media by project team members from use by a project manager for thought leadership and to deepen networks.

A role for social media in supporting communication and engagement across the project boundary is suggested by a range of authors: obtaining feedback from stakeholders (Harrin, 2010); integrating external collaborators and identifying expertise (Gimpel et al., 2014); stakeholder management (Ninan et al., 2020; Lobo and Abid, 2020; Williams et al., 2015); and branding (Ninan et al., 2019). Accordingly, four categories of activities for social media on projects are used here: project work, management activities, engagement, and external communication.

2.2.3 Impacts of Using Social Media

A survey of business executives suggested some benefits of Web 2.0 applications that are relevant for project work: increasing speed to access knowledge, reducing communication costs, increasing speed to access internal/external experts (Bughin et al., 2011). Research on the post-adoptive outcomes of using collaborative technologies found that system usage and satisfaction have been positively correlated with collaborative performance and that collaborative performance, through trust and the affective reaction called 'flow', were positively correlated with creativity (Elie-Dit-Cosaque and Pallud, 2012).

Projects require communication practices that go beyond transaction confirmation to include managing relationships, building trust, and managing stakeholder expectations. Remidez and Jones (2012) suggest that potential benefits of using social media for projects include efficiency, trust, facilitating access to tacit knowledge exchange, and spanning the project boundary. Six impacts of using social media in projects were identified by Gimpel et al. (2014): efficiency, productivity, flexibility, transparency, motivation, and trust. Benefits suggested by Van der Merwe (2016) are quick, ad-hoc information sharing, flexibility by providing information from any location, quality by fostering continuous improvement, that help to build trust and morale. Van der Merwe (2016) also highlighted a concern about the potential for information overload. Overall, previous work suggests eight benefits and one concern of using social media in project settings.

The constructs explored in this section add depth to the outline framework shown in Fig. 2.1. This new conceptual framework brings together social media and project management in a way that enables exploration of new practices and avoids giving precedence to either aspect.

2.3 Research Method and Setting

The aim of this research is to develop an understanding of the technical and human components in the use of social media for managing projects. Abductive research is apt for applying and developing new frameworks (Blomquist et al., 2010) and for moving between everyday concepts and meanings, lay accounts, and social science explanations (Mason, 2002). Qualitative methods enable experience and meaning to be uncovered and respond to calls to take seriously the lived experience of projects (Cicmil et al., 2006). Projects are context-specific and located in open systems and this is often overlooked in research design (Smyth and Morris, 2007: 423). For this research, multiple projects on a single site were chosen to reduce variation of context and organizational characteristics. Non-directive interviews were used to generate data about wide-ranging views on what was happening and to uncover new insights (Shepherd, 2015). Team interviews were used to reveal shared meaning that is an integral component for understanding social behavior (Tindale et al., 2002: 3). Consistency was optimized and bias minimized by analyzing data using a standardized process of recursive data abstraction (Jivaketu, 2015).

The projects chosen were undertaken by students as early career project managers, and their experience of using social media pre-dates their experience of formally managing projects. Social media practices among young people differ from the practices of previous generations (Prensky, 2001) and projects undertaken by students as part of their studies can be close to authentic workplace projects. To maximize authenticity, only projects that met specific criteria were included. Projects selected had to: engage external stakeholders, address a genuine problem or opportunity, expected to create artefacts to a professional standard, and have been completed with the previous year.

An initial pilot was used to test the research approach and was followed by two stages of data generation and analysis, and then a validation stage. Constructs developed in the pilot stage were subsequently refined and further new constructs were developed. Data collection and iterations of induction and deduction continued to be used until no new constructs were identified. In the final stage, a deductive approach was chosen to enable the constructs to be validated with professional project managers from a range of industries using directed interviews. The stages are summarized in Table 2.1.

Table 2.1: Summary of Data Sources.

Stage	Interview Type	Participants	No. of Interviews	No. of Direct Participants	No. of Project Participants
1- Pilot	Non-directed, team	Two teams of final-year business students undertaking consultancy projects (A, B)	2	8	9
2 (i)	Non-directed, team	Three teams of final-year business students undertaking consultancy projects (M, S, V)	3	9	9
2 (ii)	Non-directed, individual	Individual final-year business students involved with managing projects (N, T, Y)	3	3	12
2 (iii)	Non-directed, individual	Second-year students involved with managing projects (C, D, E, F, P, U)	5	6	24
3	Non-directed, pairs	Business studies alumni (X, Z)	2	4	> 40*
Validation	Directed, individual	Professional project managers (G, H, J, K, L, R)	6	6	> 60*
Totals			21	36	> 154*

* Conservative estimates of the number of people involved in the projects under discussion.

2.4 Social Media Adoption in Project Settings

A series of in-depth, un-structured interviews, lasting up to 75 minutes generated data about how early career project managers actually use social media in project settings. The data was analyzed using recursive abstraction to identify and then validate constructs for each of the four components of the conceptual framework, namely influencing factors, technology types, activities, and impacts. The constructs found for each component are discussed below.

2.4.1 Types of Social Media

A wider range of technologies was identified in the literature than participants were found to be using. Six technology types were identified by participants and ranked in order of the number of interviews in which they were mentioned as a proxy for popularity. These were: instant messaging; social networks; shared workspaces; online meetings; video and image sharing; and micro blogging. Events calendar/task scheduling tools were not explicitly mentioned in any interviews but are suggested by APM (2014) as useful for managing projects. Similarly, wikis and blogs were identified in the literature but not found in practice. Technology use changes over time and this list represents how young people were managing projects at the time the research was conducted (2013–2015). Since the research was undertaken, new apps have come to the fore and instant messaging functionality has been embedded in other software.

2.4.2 Factors Influencing Adoption and Use

Factors influencing the adoption and use varied for each type of social media. Overall, twenty-four factors were found in practice and are grouped into four types of characteristics—organizational, technological, team, and task, as shown in Table 2.2. Collaborative technologies and their use in organizations that has been reported in previous work has been driven top-down by the organization. In contrast, the subject of this research is the adoption of technologies from the personal realm for use on projects. Other literature suggesting how projects and project managers might use social media originates from the professional world. The results presented here provide the empirical data required to better understand the factors influencing social media practices.

Organizational characteristics were discussed by the participants who worked in a business setting, i.e., alumni (in stage 3) and the professional practitioners (interviewed in the validation stage). Three characteristics of the organization were identified and all three were correlated to factors identified in the literature. Security requirements and internal systems can

Table 2.2: Factors Influencing Choice and Use of Social Media (adapted from Thompson, 2017).

Factors Influencing Adoption & Use	**Type of Technology**					
	Social network	Shared workspace	Instant messaging	Online meetings	Video & image sharing	Micro blogging
Organizational characteristics						
Security requirements/policy	Y	Y	Y	Y	Y	Y
Internal systems in use/policy	Y	Y	Y	Y		
Management support	Y					
Technological characteristics						
Mobile access/accessibility	Y	Y	Y			Y
Ease and speed of use	Y	Y	Y	Y		Y
Degree of customization	Y	Y	Y			
Cost	Y	Y	Y			
Reliability		Y				
Security/privacy offered	Y	Y	Y	Y	Y	Y
Compatibility with equipment/platform		Y	Y			
Functionality						
– support for project management	Y		Y			
– document management	Y	Y				
– synchronous editing		Y				
– file size/formats supported	Y	Y	Y			
– group discussion	Y		Y			
– notifications	Y	Y	Y			
– translation	Y					
Team characteristics						
Communications preferences	Y	Y	Y	Y		
Prior technology experience	Y	Y	Y			
Social ties	Y		Y			
Team size	Y		Y	Y		
Age/generation	Y	Y	Y			Y
Experience of working together			Y			
Trust			Y			
Geographic distance	Y		Y	Y		
Job role	Y					
Task characteristics						
Extent of conversation required	Y		Y			
Size/format of file/s involved	Y	Y	Y			
Security requirements	Y	Y	Y	Y	Y	Y
Type of project/task					Y	

be viewed as technology-facilitating conditions (Brown et al., 2010), and management support was also found by Brown et al. (2010).

Following Kügler et al. (2013), characteristics of the technology refers to perceptions of using social technologies, rather than perceptions of the technologies themselves, and eight characteristics were identified in this category. The most frequently mentioned characteristics relate to perceptions of accessibility on mobile devices and the ease and speed of use. Mobile access was particularly highlighted in the data as a factor for social networks, instant messaging, and shared workspaces. Customisation, cost, and reliability were factors found in this research that were not identified in previous literature. A possible reason why these were not identified previously is that these factors may have been considered as infrastructure issues with the cost borne by the organization. Perceptions of security (APM, 2014) and compatibility (Kügler et al. 2013) are confirmed as an important consideration for use of social media in project settings. The functionality factors identified here are interpreted as forms of 'perceived usefulness' (Kügler et al., 2013; Brown et al., 2010; Venkatesh et al., 2003; Davis, 1989) and extend understanding of the role of social media in project settings.

Nine factors were identified that are characteristics of the team and team members, and five of these are considered similar to factors identified in literature. Prior technology experience (Giltenane, 2016; Kügler et al., 2013; Brown et al., 2010; Venkatesh et al., 2003), experience of working together (Brown et al., 2010), social ties (Kügler et al., 2013; Brown et al., 2010), age (Giltenane, 2016; Venkatesh et al., 2003) and trust (Giltenane, 2016; Kügler et al., 2013) are confirmed as factors influencing the adoption and use of social media in project settings. New characteristics of the team identified are job role, team size, geographic distance, and communications preferences, and the role of these characteristics could be investigated in further research. Further research is also required to understand the influence of the four task characteristics found because these differ from the characteristics of the project or task identified in literature.

2.4.3 Activities Supported or Enabled by Social Media

Social media were used in four types of activities in project settings—project work, management activities, engagement activities, and external communication. Twenty-one communication activities and the types of technology used for each are recorded in Table 2.3. Combining literature from organizational learning, social communication, and social media use in project management suggest three levels or zones of project communication activities. It should be noted that social media use for project work and management activities are combined in the conceptual level of project action and learning.

Table 2.3: How Social Technologies were Used (adapted from Thompson, 2017).

Activities	Type of Technology					
	Social network	Shared workspace	Instant messaging	Online meetings	Video & image sharing	Micro blogging
Project work						
Storing information	Y	Y				
Sharing information	Y	Y	Y	Y		
Sharing work	Y	Y	Y	Y		
Management activities						
Assigning tasks and recording allocation	Y		Y	Y		
Checking work progress	Y		Y	Y		
Reporting task status	Y	Y	Y			
Decision making	Y		Y	Y		
Project diary	Y	Y			Y	
Change management	Y	Y	Y	Y		
Capturing lessons learned		Y			Y	
Engagement activities						
Organizing and support for meetings	Y	Y	Y	Y		
Conducting meetings			Y	Y		
Informal information sharing	Y		Y	Y		
Requesting and providing feedback	Y		Y	Y		
Reminders and prompts	Y		Y	Y		
Solving problems and discussion	Y		Y	Y		
Brainstorming and sharing ideas	Y		Y	Y	Y	
Keeping in contact/helping others	Y	Y	Y	Y		
External communication						
Distributing information externally	Y	Y		Y	Y	Y
Gathering external information	Y			Y		Y
Contacting external people	Y			Y		

Three communication activities were recorded as project work and correspond to activities identified in the literature. Of the seven communication activities recorded as management activities, five were identified in the literature and two are new ways that social media can be used—change management and capturing lessons learned (Thompson, 2017). Eight activities were identified as communication activities related to engaging team members and of these, two were not found in previous literature—'reminders and prompts' and 'keeping in contact and helping others' (Thompson, 2017). Three external communication activities were found where social media had been used. Integrating external collaborators into project using social media was suggested in the literature (Gimpel et al., 2014; Harrin, 2010a) and is confirmed here, although the specific technologies suggested are different.

2.4.4 Impacts of Using Social Media in Project Settings

Perceptions of the impacts of using social media in project settings were recorded and classified as either benefits or concerns, depending on how participants described each construct. The constructs were grouped into seven different categories and recorded for each of the technology types, as shown in Table 2.4 and Table 2.5.

Efficiency and productivity benefits were identified in the literature (Gimpel et al., 2014; Van der Merwe, 2016; Remidez and Jones, 2012; Bughin et al., 2011). Five different ways in which social media is perceived as a benefit and four concerns were recorded here and, in this way, understanding of the impacts on efficiency is extended. Increased opportunities for feedback using social media improves the quality of work and was recorded and supports the suggestion made by Van der Merwe (2016). Nine ways in which social media are perceived to improve information management and one concern were recorded here, and this level of detail was not found in the literature. Three different perceptions of flexibility and three of transparency as benefits of using social media were recorded. Both flexibility and transparency were identified by Gimpel et al. (2014) and this work extended understanding of what these terms mean in project settings. Two perceptions of how creativity is a benefit of using social media were found, thereby confirming and extending understanding of creativity suggested by Elie-Dit-Cosaque and Pallud (2012). The final category of impacts were perceptions collectively termed 'emotional impacts' (Thompson, 2017) where six types of beneficial impact and two concerns were recorded. Motivation and trust were identified in literature (Van der Merwe, 2016; Gimpel et al., 2014) and confirmed here; specifically, trust was perceived to develop faster when social media were used. Perceptions of four new beneficial impacts and two concerns were identified, adding to the literature.

Table 2.4: Benefits of Using Social Media in Project Settings (adapted from Thompson, 2017).

Benefit	**Type of Technology**					
	Social network	Shared workspace	Instant messaging	Online meetings	Video & image sharing	Micro blogging
Efficiency						
Saves time/immediacy	Y	Y	Y	Y		Y
Reduces need to meet face-to-face	Y		Y	Y		
Synchronous document editing		Y				
Driving project forward	Y		Y	Y		
Convenient	Y	Y				
Quality of work						
Feedback improves quality	Y		Y	Y		
Information management						
Eliminates need to send	Y	Y				
Easier information sharing	Y	Y	Y		Y	Y
Faster information sharing	Y		Y	Y		
Saves duplication		Y				
Facilitates informal info sharing	Y		Y	Y		
Extends information sharing	Y				Y	
Easy to find information	Y					
Information not missed	Y		Y			
Flexibility						
Ability to work in own time	Y	Y	Y	Y		
Ability to work at any location	Y	Y	Y	Y		
Dynamic task allocation	Y		Y			
Transparency						
All have the same information	Y	Y	Y		Y	
Visibility of work and status	Y		Y			
Accountability	Y		Y			
Creativity						
Facilitates sharing of ideas	Y		Y	Y	Y	
Stimulates thinking					Y	
Emotional impacts						
Encourages participation/motivation	Y		Y	Y		
Increases focus	Y		Y			
Enjoyment				Y	Y	
Feeling connected/engaged	Y		Y	Y	Y	
Faster trust			Y	Y		
Encourages sensitivity	Y		Y			

Table 2.5: Concerns about Using Social Media in Project Settings (adapted from Thompson, 2017).

Concern	Technologies					
	Social network	Shared workspace	Instant messaging	Online meetings	Video & image sharing	Micro blogging
Efficiency						
Less concise/less formal	Y		Y			
Tasks ignored	Y		Y			
Losing track of work			Y			
Distracting			Y			
Information management						
Information loss	Y	Y	Y	Y		
Emotional impacts						
Fear of overload	Y		Y			
Fear of letting team down/mistakes	Y		Y			

2.5 Knowledge Management in Projects

New knowledge is created during a project and managing knowledge is essential for project success. Knowledge transformation involves the four processes of the SECI model (Nonaka and Konno, 1998) and all are considered essential for developing new knowledge and yet project management processes emphasize explicit knowledge, with little or no attention to tacit and causal knowledge (Reich and Wee, 2006). Traditionally, no project management processes address the socialization and internalization processes, leaving gaps in knowledge management, as illustrated in Fig. 2.2.

For each of the activities recorded in this research, the knowledge involved was used to map the activities against the four knowledge processes of the SECI model, as illustrated in Table 2.6.

Communication activities using social media are identified for all four knowledge processes. Engagement activities, in particular, address the two knowledge processes not addressed by traditional project management processes—socialization and internalization. In this way, using social media in project settings is shown to address the knowledge gaps of traditional project management, as illustrated in Fig. 2.3.

All the engagement activities, with the exception of conducting meetings, involved use of a social network and instant messaging. Instant messaging was perceived to be useful for contacting team members, for purposes including organizing meetings and informal knowledge sharing,

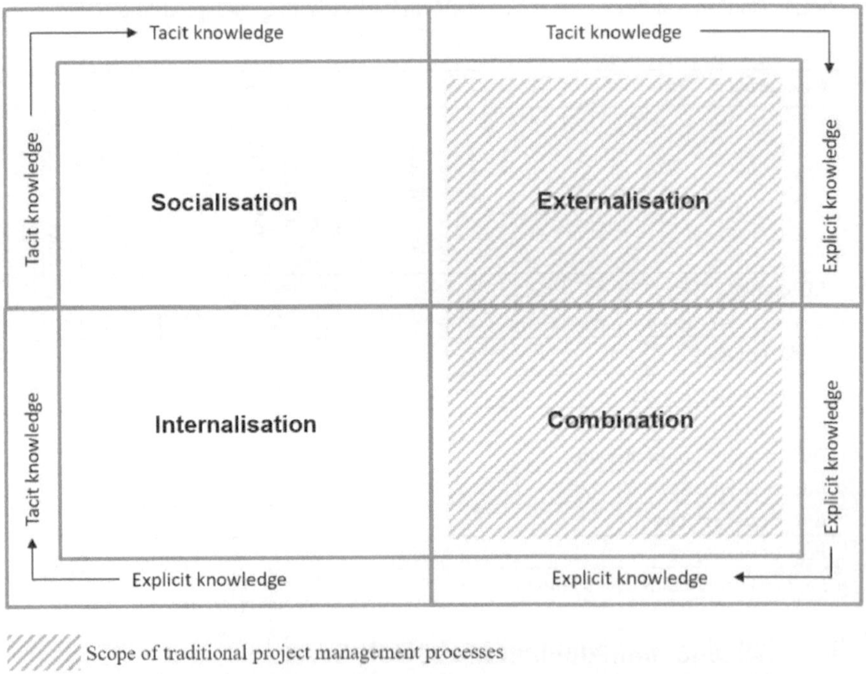

Fig. 2.2: Traditional project management processes in the SECI model of knowledge management (using Thompson, 2017; Reich and Wee, 2007; Nonaka and Konno, 1998). (Alt Text: SECI model showing the transformation of tacit and explicit knowledge through four processes)

and notifications kept team members up to date with information and changes. Perhaps, unsurprisingly, one perceived benefit of social media was a reduction in the need for face-to-face meetings. In addition, social media were used to keep team members engaged with the project in between face-to-face meetings, as explained by members of Team B:

> '… we use it as a way to complement the meetings we have and kind of fill in the gaps…' (B)

Messaging was particularly important for filling in the gaps, both in between meetings and during meetings. Dennis et al. (2010) found instant messaging enabled 'invisible whispering' that alters socio-spatial temporal boundaries. This is consistent with use of instant messaging being used concurrently with online meetings, as explained by Project Manager G:

> '… actually, you can actually have a sub-conversation with others to say when, when we come to section three in the next piece will you, can you pick up that for me please, can you run that piece…. by using online and conferencing, use the instant messaging to act, at the same time.' (G)

Table 2.6: Mapping of Project Management Activities to Knowledge Transformation Processes (Thompson, 2017).

Activities	Knowledge Input	Knowledge Output	Knowledge Management Process
Project work			
Storing information	Explicit	Explicit	Combination
Sharing information	Explicit	Explicit	Combination
Sharing work	Explicit/ Tacit	Explicit/ Tacit	Combination/Externalization/ Internalization/Socialization
Management activities			
Assigning tasks & recording allocation	Tacit	Explicit	Externalization
Checking work progress	Tacit	Explicit	Externalization
Reporting task status	Tacit	Explicit	Externalization
Decision making	Tacit	Explicit	Externalization
Project diary	Tacit	Explicit	Externalization
Change management	Tacit	Explicit	Externalization
Capturing lessons learnt	Tacit	Explicit/ Tacit	Externalization/Socialization
Engagement activities			
Organizing and support for meetings	Tacit	Explicit/ Tacit	Externalization/Socialization
Conducting meetings	Tacit	Tacit	Socialization
Informal information sharing	Tacit	Tacit	Socialization
Requesting and providing feedback	Tacit	Tacit/ Explicit	Socialization/Externalization
Reminders and prompts	Tacit	Explicit	Socialization
Solving problems and discussion	Explicit/ Tacit	Tacit/ Explicit	Externalization/ Combination/Internalization/ Socialization
Brainstorming and sharing ideas	Tacit	Tacit	Socialization
Keeping in contact/ helping others	Tacit	Tacit	Socialization
External communication			
Distributing information externally	Explicit	Explicit	Combination
Gathering external information	Explicit/ Tacit	Explicit	Combination/Internalization
Contacting external people	Tacit	Tacit/ Explicit	Socialization/Externalization

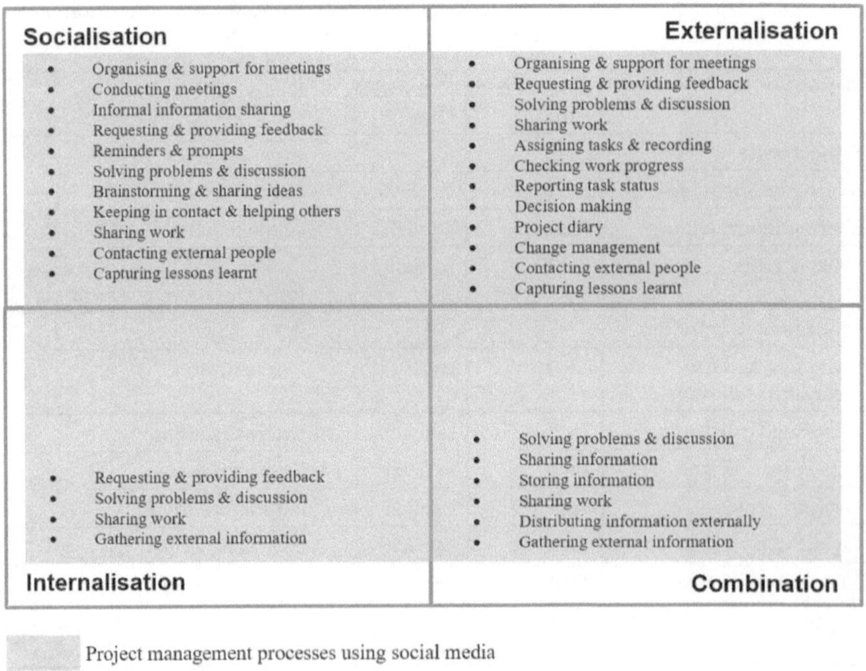

Fig. 2.3: Activities involving social media in knowledge transformation (Thompson, 2017). (Alt Text: SECI model with the mapping of project management activities that involve social media)

Social media transforms content as well as frequency of communication. The dominant paradigm for communication in project management traditionally has been documents and reports. Use of social media to share images and videos with or without textual narrative is commonplace, and links to videos, images or other website content are easily shared on social networks. Communication content is no longer restricted to narrative-text in documents and voice in phone calls. Content in the form of graphics, audio, images, and videos can easily be created and shared using links to web-based material over social networks, instant messages, and other social technologies. Online meetings enable audio and visual content to be created and shared in real time. This work records illustrations of social media use that increase communication amongst team members in ways that facilitate all four processes that are essential for knowledge creation. In this way, social media are capable of addressing some limitations of traditional approaches to project management.

2.6 Communication Across the Project Boundary

Poor communication with external stakeholders has been highlighted as an issue for projects (Müller, 2003; Partington, 1997). Conventional approaches to project management include some attention to communication outwardly to inform external stakeholders about the project. Little if any attention is given to actively seeking information about changes in the external environment that could affect implementation or integration of the outputs into business-as-usual, and even less looking for new opportunities to improve the project outcomes, benefits, or value economically, socially, or environmentally. Agile approaches address some shortcomings of linear approaches typically by co-locating the team and other stakeholders, holding regular meetings, using short timeframes, etc. But agile approaches emphasize delivery of the product or service and strengthen perceptions of the project boundary. A project boundary is drawn to help manage scope and is defined early in a project. Perceptions of a strong boundary to the project can mean that opportunities for responding to change or improving the outcomes are missed. For example, if the project is defined as software development, then software will be developed that may not be underpinned by good understandings of the behavioral changes required to ensure success. For example, some countries approached the development of mobile apps for tracking to limit the spread of COVID-19 without adequate understanding of the social systems required for large-scale adoption. Other examples can be seen in major road infrastructure projects that fail to fully understand the ecological systems they impact and are delayed by environmental protestors. Projects that lack effective communication across the project boundary can become disconnected from the external world and are less likely to successfully deliver the expected outcomes.

This research found use of social media to support and enhance communication across the project boundary. Use of social networks and online meetings were recorded for all three activities in the external communication category, namely, distributing and gathering information, and contacting external people. In addition, some examples were recorded of sharing information externally by using shared workspaces, video and image sharing, and micro-blogging, as well as gathering information and feedback using micro-blogs, as explained by one participant:

> 'So, I might use um hash tags against comms campaigns and then search for those hash tags and then analyze sort of who's been commenting, who's been liking those posts ... that's to post top tips, and um end user support on everything to do with IT products and services. ... we can measure is the amount of likes and ... comments' (Z)

Social media transforms not just the key components of communication identified in literature of medium, frequency, and content (Müller, 2003), but also can cross the project boundary by providing access to experts that are external to the team, as explained here:

> '... it opens the door to an expert you might not even know is there ... it's still opening up the expertise to those two hundred and thirty people within IT that you might not talk to on a daily basis.' (Z)

As the choices for project communication increase, so too do expectations. One project manager commented on unsuccessful use of a blog and stakeholders' expectations for using channels already familiar to them:

> 'We've consistently run a blog on quite a significant project ... We still get non-stop complaints about you're not communicating enough, when we point people at the blog it's kind of, I can't be bothered to log in and read that what I want is an email.... they expect the information to come to them.' (J)

In a conventional approach, project communication tends to primarly flow in one direction, namely outwards from the project. Viewing a project as a learning organization highlights a requirement for information to flow into a project from the environment. This research provides evidence of practitioners using social media to exchange information across the boundary in both directions, more easily, faster, and potentially with a wider range of stakeholders but there are some caveats about the choices to be made.

2.7 Conclusion

In this chapter I set out to explain the potential of social media to enhance project communication in ways that improve the likelihood of delivering projects successfully. Management practices in many organizations are shaped by a deterministic view of projects and therefore participants were chosen whose experience of social media pre-dates their experience of formal project management. I began by developing a socially constructed re-conceptualization that challenges objectivist approaches by bringing human interaction and communication to the fore. The experiences of early career project managers were explored because young people tend to be early adopters of new technology and may use social media differently from previous generations (Prensky, 2001).

The research provides empirical evidence for how social media are actually used in project settings. This understanding of project communication practices will help engage both internal and external stakeholders. Theoretically, firstly I highlight the role of social media in connecting a project with its external environment and increasing visibility of the project to stakeholders. Secondly, I indicate the ways in which social media facilitate learning about the project context and support change management. Third, I explain the contribution social media makes to knowledge creation through supporting internalization and socialization. There are multiple contributions to practice from this work. The factors influencing use of social media are documented for six types of technology in project settings. Next, I record communication activities supported by use of social media for project activities in three categories and highlight some activities that extend understanding of project management. This research also identifies new understandings of how social media can enable flexibility and creativity in teamwork. Finally, social media can be used to increase project communication in ways that can help a project respond to change. Increasing external communication in ways that increase knowledge of the project environment may be viewed as increasing disruption to the project, but also may mean better understanding of the requirements for integration of the project outputs, thereby improving the likelihood of successfully achieving the project outcomes.

One limitation of the research is the self-selection of the participants and the possibility that only those whose experience was largely positive volunteered to take part. Use of technology changes over time and new technology emerged during the time of the research. As social media savvy young people make up a growing proportion of the workforce, practices in many aspects of business are changing. Further research would be useful to extend the understandings presented here by revealing more recent technology, influencing factors, activities, and perceptions of impacts that may now be relevant in commercial settings.

The majority of projects discussed in this work were undertaken by teams of three to five people in an HEI setting with no, or very little, monetary budget. Cost, time, and quality are widely considered important project constraints to be managed. Most student projects have little or no actual cost to be managed, and therefore are different to most commercial projects. The impact of a zero budget is unclear and managing a project with no budget might be more or less challenging than managing a project with a budget. Most student projects have a time constraint with no tolerance, i.e., an immovable deadline. Not all projects have a fixed timeframe and, again this may make a project more or less challenging to manage. Quality is the third project constraint commonly managed. Where student projects are formally assessed, quality is a concern because one focus for students is the grade they will be awarded for the project. Therefore, although most projects took place in an HEI setting, the 'triple constraint' is applicable.

Validation involved discussing the results with professional practitioners from a wide range of organizational settings and therefore the findings are likely to be applicable to projects in a wide range of industries.

References

Ajzen, I. (1991). The theory of planned behavior, *Organization Behavior and Human Decision Process*, **60**: 179–211.

APM. (2014). APM Web Briefing—Social Media in Project Management, Is It a Waste of Time? Association for Project Management: Buckinghamshire. Available online: https://www.apm.org.uk/sites/default/files/apm%20web%20briefing%20-%20social%20media%20in%20project%20management,%20is%20it%20a%20waste%20of%20time_0.pdf [accessed on 9 November 2021].

Blomquist, T., Hällgren, M., Nilsson, A., and Söderholm, A. (2010). Project-as-practice: In search of project management research that matters, *Project Management Journal*, **41**(1): 5–16.

Bok, H.S., Kankanhalli, A., Raman, K.S., and Sambamurthy, V. (2012). Revisiting media choice: A behavioral decision-making perspective, *International Journal of e-Collaboration*, **8**(3): 19–35.

Boulding, K. (1956). General systems theory: The skeleton of science, *Management Science*, 2,3 (Apr. 1956). Available online http://panarchy.org/boulding/systems.1956.html [accessed on 9 November 2021].

Brown, S.A., Dennis, A.R., and Venkatesh, V. (2010). Predicting collaboration technology use: Integrating technology adoption and collaboration research, *Journal of Management Information Systems*, **27**(2): 9–53.

Bughin, J., Chui, M., and Pollack, L. (2013). Organizing for change through social technologies, *McKinsey and Company High Tech Survey*, McKinsey and Company. Available online: http://www.mckinsey.com/industries/high-tech/our-insights/organizing-for-change-through-social-technologies-mckinsey-global-survey-results [accessed on 9 November 2021].

Cicmil, S., Williams, T., Thomas, J., and Hodgson, D. (2006). Rethinking project management: Researching the actuality of projects, *International Journal of Project Management*, **24**: 675–686.

Daft, R.L., and Weick, K.E. (1984). Towards a model of organizations as interpretation systems, *Academy of Management Review*, **9**(2): 284–295.

Davis, F.D. (1989). Perceived usefulness, perceived ease of use, and user acceptance of information technology, *MIS Quarterly*, September 1989, pp. 319–340.

Dennis, A.R., Rennecker, J.A., and Hansen, S. (2010). Invisible whispering: Restructuring collaborative decision making with instant messaging, *Decision Sciences*, **41**(4): 845–886.

Elie-Dit-Cosaque, C., and Pallud, J. (2012). Understanding collaborative technologies adoptive and post-adoptive outcomes: A longitudinal study, association for information systems, *AMCIS 2012 Proceedings Paper*, 29.

Giltenane, J. (2016). Investigating the intention to use social media tools within virtual project teams. pp. 83–105. *In:* Silvius, G. (Ed.). *Strategic Integration of Social Media into Project Management Practice*, Hershey: IGI Global.

Gimpel, H., Floesser, P., Caton, S., and Schaefer, J. (2014). Social collaboration in project work: An exploratory empirical analysis of benefits and technologies, *SSRN*. Available at: https://ssrn.com/abstract=2418447 [accessed on 9 November 2021].

Goodhue, D.L., and Thompson, R.L. (1995). Task-technology fit and individual performance, *MIS Quarterly*, pp. 213 –236.

Grudin, J. (2010). CSCW: Time passed, tempest, and time past. *Interactions*, **17**(4): 38–40.

Harrin, E. (2010). *Social Media for Project Managers*, Pennsylvania: PMI.

Hollocks, B.W. (1994). *The Convergence of Parallel Lines*, Inaugural lecture of Intel Corporation Chair in Computer Supported Co-operation, 11th May 1994, Bournemouth University, Department of Management Systems.

Jivaketu, P. (2015). *An Evaluation of Teacher Development in Using Technology during the First Decade of Thai Education Reform 1999–2009*, Boston University. Available online: https://open.bu.edu/pdfpreview/bitstream/handle/2144/15197/Jivaketu_bu_0017E_10994.pdf?sequence=1 [accessed on 9 November 2021].

Kügler, M., Smolnik, S., and Raeth, P. (2013). Determining the factors influencing enterprise social software usage: Development of a measurement instrument for empirical assessment, *46th International Conference on System Sciences*, IEEE Computer Society, Hawaii, 3635–3644.

Lobo, S., and Abid, A.F. (2020). The role of social media in intrastakeholder strategies to influence decision making in a UK infrastructure megaproject: Crossrail 2, *Project Management Journal*, **51**(1): 96–119.

Mason, J. (2002). *Qualitative Researching*, 2nd ed., London: Sage.

McGuire, W.J. (1969). The nature of attitudes and attitude change. *In*: Lindzey, G., and Aronson, E. (Eds.). *The Handbook of Social Psychology*. 3: 136–314. Reading, MA: Addison-Wesley.

Müller, R. (2003). Determinants for external communications of IT project managers, *International Journal of Project Management*, 21: 345–354.

Ninan, J., Clegg, S., and Mahalingam, A. (2019). Branding and governmentality for infrastructure megaprojects: The role of social media, *International Journal of Project Management*, **37**(1): 59–72.

Ninan, J., Mahalingam, A., Clegg, S., and Sankaran, S. (2020). ICT for external stakeholder management: Socio-materiality from a power perspective, *Construction Management and Economics*, **38**(9): 840–855.

Nonaka, I., and Konno, N. (1998). The concept of "Ba": Building a foundation for knowledge creation, *California Management Review*, **40**(3): 40–54.

Partington, D. (1997). Management Processes in Projects of Organizational Change: Case Studies from Four Industries, UK: Cranfield University.

Prensky, M. (2001). Digital Natives, Digital Immigrants, *On the Horizon*, vol. 9, No. 5. Available online: http://marcprensky.com/writing/Prensky%20-%20Digital%20Natives,%20Digital%20Immigrants%20-%20Part1.pdf [accessed on 9 November 2021].

Reich, B.H., and Wee, S.Y. (2006). Searching for knowledge in the PMBOK® Guide, *Project Management Journal*, **37**(2): 11–26.

Remidez, H., and Jones, N.B. (2012). Developing a model for social media in project management communications, *International Journal of Business and Social Science*, **3**(3): 33–36.

Shepherd, M. (2015). Interview methods for project management research. pp. 185–202. *In:* Pasian, B. (Ed.). *Designs, Methods and Practices for Research of Project Management*, Farnham: Gower.

Silvius, C.M., and Silvius, G. (2016). 'Smart' project management: smartphone 'apps' for project management. pp. 179–192. *In*: Silvius, G. (ed.). *Strategic Integration of Social Media into Project Management Practice*. Hershey: IGI Global.

Smyth, H.J., and Morris, P. (2007). An epistemological evaluation of research into projects and their management: Methodological issues, *International Journal of Project Management*, **25**(2007): 423–436.

Taylor, P. (2016). The Social Project Manager: Balancing Collaboration with Centralized Control in a Project Driven World, *PM World Journal*, vol. 5, Issue 5, pp. 1–6. Available online: http://pmworldlibrary.net/wp-content/uploads/2016/05/pmwj46-May2016-Taylor-the-social-project-manager-Series-Article.pdf [accessed on 9 November 2021].

Thompson, K.M. (2017). *A Framework for using Social Media in the Practice of Project Management*, PhD Thesis, UK: Bournemouth University.

Tindale, R.S., Meisenhelder, H.M., Dykema-Engblade, A.A., and Hogg, M.A. (2002). Shared cognition in small groups. *In:* Hogg, M.A. and R.S. Tindale (Eds.). *Blackwell Handbook of Social Psychology: Group Process*, Blackwell Publishing.

Turban, E., Liang, T.P., and Wu, S.P.J. (2011). A framework for adopting collaboration 2.0 Tools for virtual group decision making, *Group Decision and Negotiation*, 20: 137–154.

Turner, J.R. (1999). *The Handbook of Project-based Management: Improving the Process for Achieving Strategic Objectives*, London: McGraw-Hill.

Turner, J.R. (2014). *Gower Handbook of Project Management*, 5th ed., Gower e-book.

Van der Merwe, L. (2016). Social media use within project teams: Practical applications of social media on projects. pp. 139–159. *In:* Silvius, G. (Ed.). *Strategic Integration of Social Media into Project Management Practice*, Hershey: IGI Global.

Venkatesh, V., Morris, M.G., Davis, G.B., and Davis, F.D. (2003). User acceptance of information technology: Toward a unified view, *MIS Quarterly*, 27(3): 425–478.

Williams, N.L., Ferdinand, N., and Pasian, B. (2015). Online stakeholder interactions in the early stage of a megaproject, *Project Management Journal*, **46**(6): 92–110.

Zhao, Y., Tang, J., and Zhu, Q. (2013). Conceptualising perceived affordances in social media interaction design, *Aslib Journal of Information Management*, **65**(3): 289–302.

3

The use of Social Media for Communication and Knowledge Sharing in the Home-Office Work Period

Senior Project Worker's Perspective

Marcirio Silveira Chaves,[1,*] *Giuliana Veronese*[2] and *Roberto Winter*[3]

3.1 Introduction

Information Technology is changing the way all interactions are made (Ou et al., 2014), including how employees of contemporary organizations engage, making them less constrained by space and time (Qureshi et al., 2018). Research on social media (SM), including social networks (e.g., Facebook and LinkedIn), enterprise social media platforms (e.g., Yammer), enterprise instant messaging (e.g., Facebook Messenger), mobile social media (e.g., WhatsApp and WeChat), and Web 2.0 applications (e.g., wiki and blogs) has increasingly gained the attention of management (Kane, 2017; Sun, 2019) and project management (Rosa et al., 2016; Zhang et al., 2018) communities.

[1] Pontifical Catholic University of Rio Grande do Sul, Brazil.
[2] Stefanini EMEA, Brazil.
[3] Tata Consultancy Services, Brazil.
* Corresponding author: mschaves@gmail.com

The use of social media tools pervades our society and organizational settings in such a way that their benefits and challenges can no longer be ignored (Ahmed et al., 2019). The imperative necessity to use collaborative solutions during the COVID-19 pandemic has made existing barriers disappear, and minds have opened to the advantages of social media (Ozguler, 2020). Social media enables more flexible and socially distanced or remote work. Working remotely has been a frequent research topic across such diverse areas, as information systems, human resources, management, computer science, psychology, and other disciplines for the past two decades (Hacker et al., 2019). The COVID-19 pandemic brings a new urgency to understand better how remote teams perform, since working from home is no longer a privilege but a necessity (Banjo et al., 2020).

Although there is currently a high growth in social media that supports collaboration, knowledge sharing, and communication, the use of these tools in project management remains limited and has difficulty in keeping apace with the development of such tools (Nach, 2016). In addition, the physical distance between stakeholders brings more significant challenges to the possibilities of contributing to the mutual benefits of an IT project, as it generates problems of communication (Henderson et al., 2016) and knowledge sharing (Naeem, 2019).

Communication has been reported as one of the leading causes of project failure (Dwivedi et al., 2015). Poor communication affects IT projects by incorporating significant risks, such as delay in shipments of sprints, over-budgeted shipments, lack of motivation, unclear goals, and friction among team members (Qureshi et al., 2018). When new information comes up, and the information workers do not know whom to inform, they may not inform anyone or send information to unnecessary people, leading to conflicts (Majdenbaum and Chaves, 2020).

Knowledge sharing can improve the quality of work within a project and between projects (Nesheim and Hunskaar, 2015). Although leaders, consultants, and researchers are increasingly trying to intensify the adoption of social media tools to support knowledge-sharing practices, it is usually a complex and complicated endeavor (Naeem, 2019). Naeem (2019) points out that social media has limitations as technological support to enhance knowledge sharing, like fear of losing power, lack of intention to share knowledge, lower level of motivation, and resistance to technology. Understanding how social media supports knowledge sharing in remote working is one of the challenges discussed in this chapter.

Considering that the adoption or use of social media tools in projects is still in its infancy (Ikemoto et al., 2017; Kanagarajoo et al., 2019) and the existing body of research lacks empirical evidence of social media adoption in project management practices (Delerue and Cronje, 2017), this chapter tries to understand how senior project management practitioners are using social media to communicate and share knowledge during the pandemic period. This chapter also seeks to extend the research on communication and knowledge sharing in IT projects, from software development to projects including infrastructure, service desk, and field service, using interviews with senior practitioners working in this kind of project (PMI, 2013, 2020, 2021).

3.2 Theoretical Background

3.2.1 Communication in IT Projects

In this chapter, we adapt the communication concept proposed by Majdenbaum and Chaves (2020). Communication is the degree to which internal stakeholders as product owners (PO), scrum masters (SM), project managers (PM), and the development team create and share information effectively during the IT project. Majdenbaum and Chaves (2020) restrict the concept to software development projects while we extend it to IT projects, including infrastructure, service desk, and field service. Internal communication in IT projects is among the three most significant critical success factors within this communication context. The constant interaction between people, organization, and the environment, frequent in IT projects, often requires excellent communication (Iriarte and Bayona, 2020).

The acceleration of innovation in IT and changes in the corporate world have made working together, in virtual teams, a natural change in the way organizations work (Nyström and Asproth, 2013), bringing even more significant challenges to communication. The distribution of project teams can also make it easier to employ more people as there are no space restrictions. However, these benefits come with increased risks due to a lack of face-to-face interaction, resulting in a lack of trust and ineffective communication leading to difficulty in cooperation.

Among the factors that contribute to the achievement of project performance, communication has a prominent role. A structured communication model influences project performance. For an efficient flow of information, it is suggested that such a model should play the role of mediator. Thus, communication based on a model, producing improved information flow would likely improve project performance, affecting project quality and cost. It is advisable that project workers should try to promote an efficient flow of information and, with this, through a practical method, the effective management of the flow of information. Collaborative work environments can also be promoted when project professionals want to share information (Cheung et al., 2013).

Two types of communication should be considered to evaluate the effectiveness of communication in a project. Formal communication refers to the communication style which is planned, organized, and allowed by management and has rules, standards, and processes (Price, 1997). Informal communication is face-to-face conversations in non-work settings, such as after-work meetings, lunches, and parties (Park and Lee, 2014). Currently, collaborative tools, such as Zoom, Teams, and Google Meet have mediated this face-to-face communication. The use of social media for communication has become commonplace in geographically dispersed and open-source software projects. The use of SM intensifies informal communication between team members (Lanubile et al., 2010; Zhang et al., 2018).

While the internet and e-mail used to be the primary means of communication between teams, new technologies and applications are constantly emerging and these present opportunities to improve project management. More interactive communication tools have emerged in recent years, such as social media, widely used for personal purposes (Rimkuniene and Zinkeviciute, 2014). However, little has been studied regarding their ability to promote greater communication and knowledge sharing among project teams.

Communication is a fundamental competency in organizational project and programme management that drives the project team to meet project goals when appropriately applied. Project leaders and stakeholders need to understand the importance of this essential component, else the project outcomes can be jeopardized, and budgets can incur unnecessary risk. In recent years, the Project Management Institute (PMI) research group has published studies (PMI, 2013, 2020, 2021) presenting the importance of conducting a proper communication plan to maximize results and minimize risks. In addition to this, our IT workplace has been changing to a more diverse and geographically dispersed environment.

Many organizations realize the value of different perspectives that come with a diverse workgroup. Diversity can lead to more ideas and higher levels of creativity, resulting in better outcomes and innovative products and services. However, communication within a diverse workgroup can create complex and challenging situations. People with varying cultures and experiences have different meanings and contexts for words and phrases. Social media has been applied to improve communication and knowledge sharing; therefore, it is essential to review the communication methods and how the new technology has affected them.

Active communication includes meetings, telephone calls, and video conferences. Active communications form an essential part of project communication but both scheduling and the activities themselves can be time-consuming (in scheduling and the activities themselves). Active communication is best used at the onset of a project and in key phases within the project. It has been observed that specific agile methods, such as XP, instead of documentation, use active and continuous communication between team members, where team members and customers communicate continuously to arrive at problem resolution. The client actively participates by writing stories, for example. Good management and communication skills are essential for keeping team members together and on track and dealing with different stakeholders with interests (Anwer et al., 2017). Communication skills have increased, and their role is important in agile project development methods, with information sharing. Thus, effective, continuous, and active communication between project members and the client is among the project's success factors (Taibi et al., 2018).

On the other hand, passive communication allows people to process information according to their schedules. It is best used to monitor progress, ask questions of team members, and inform stakeholders of the project status. Passive communication can be divided into two different methods (PMI, 2017):

- *Push communication*. Communication is sent to specific recipients who need to receive the information. This procedure ensures that the information is distributed but does not ensure that it reached or was understood by the intended audience. Push communications include ledgers, memos, reports, e-mails, faxes, voice mails, blogs, and press releases.
- *Pull communication*. This is used for vast volumes of information or very large audiences and requires the recipients to access the communication content at their discretion. These methods include intranet sites, e-learning, lessons learned databases, and knowledge repositories.

3.2.2 Knowledge Sharing

Knowledge sharing is a strategic key to producing sustained long-term competitive advantage. It is the ability of people and organizations to disseminate expertise, insights, and best practices, increasing team productivity and empowering employees to do their jobs effectively and efficiently. Knowledge sharing is at the core of knowledge management processes. This section presents an overview of knowledge management processes and techniques to understand how social media can support these processes.

Knowledge management processes involve acquiring, sharing, and transferring knowledge between project managers, team users, top management, and other stakeholders (Foote and Halawi, 2018). Such human interactions can yield benefits as preventing mistakes from being repeated, reducing the loss of know-how, and leveraging existing experience and knowledge (Chaves et al., 2018). Sharing is the most vital among knowledge management process since most initiatives depend upon it (Anwar et al., 2019; Krumova and Milanezi, 2015).

Knowledge can be classified as follows: first, explicit knowledge refers to the visible information that can be readily articulated, codified, stored, and accessed. It may be available in videos, reports, technical specifications, policies, patents, and literature; second, tacit knowledge is the knowledge present in peoples' minds and could result from their experiences and personal values. This makes it less easy to share and therefore a knowledge management process is needed to transform tacit into explicit knowledge to benefit from the former.

Knowledge management processes do not have a limited scope, and their main objective is to improve the organization's efficiency. It ensures managing knowledge through an organized process for identifying, registering, storing, sharing, applying and renewing knowledge to create value. These processes constantly require the development of new skills and further improvement of existing capabilities. Figure 3.1 presents the knowledge management processes as continuous and systematic, involving specific processes, techniques, and tools. These processes should be standardized to maximize benefits and gain an edge in a dynamic economy. The main knowledge processes are:

Identifying knowledge: Every organisation can have multiple knowledge sources. The knowledge identification process refers to understanding what the knowledge flows are inside an organization. The primary sources

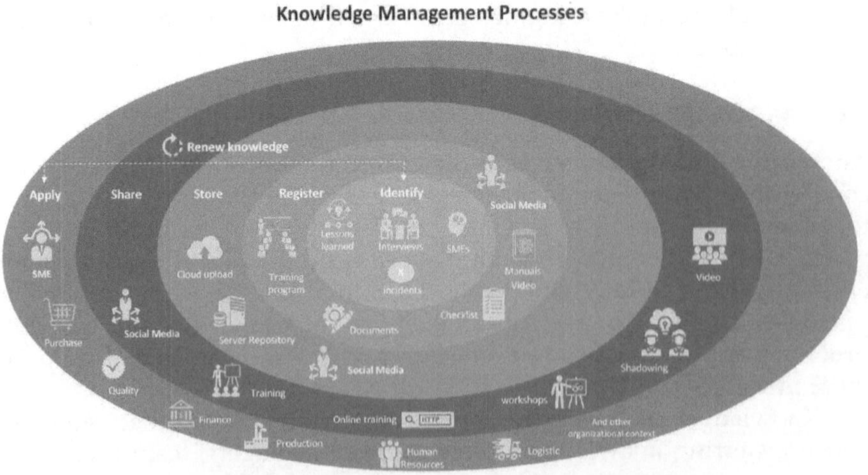

Fig. 3.1: A mapping of the knowledge management processes.
(Alt Text: A framework showing knowledge management processes such as identify, register, store, share and apply)

of knowledge are usually products and services acquisition, market trends, project lessons learned sessions, root causes analysis of incidents, problem management process, and interviews to map the knowledge with subject matter experts, project team, and others.

Registering knowledge: Collecting and registering knowledge is an essential part of the knowledge management process. If incorrect data is collected, the knowledge generated from it will not be accurate and can affect critical organizational decisions. Therefore, a formal protocol defining how the knowledge is collected, registered and who is responsible for validating it is crucial. The most common formats are videos, manuals, blogs, reports, and training programmes.

Storing knowledge: The storage process defines how the knowledge will be preserved. Knowledge can be stored in traditional shared repository structures like computer databases or modern tools like social media. There is a vast sea of unorganized knowledge present in every organization. To better use this knowledge in the future, it is necessary to store and organize it in an orderly way. The organizational structure could be increased and accessing knowledge could be easier if a system is created that is well categorized and excellently mapped.

Sharing knowledge: Knowledge distribution ensures that the information reaches the relevant people who can learn and apply that specific content. Usually, the main methods to share knowledge in organizations are notifications through e-mail and other communication tools, training sessions, shadowing activities, and workshops.

Applying knowledge: Knowledge application requires embedding knowledge in the procedures and practices of the organization. A learning organization creates an ideal environment for applying knowledge in different organizational contexts.

Reviewing knowledge: Reviewing and renewing knowledge are crucial processes to examine knowledge relevance, accuracy, and practicality. These activities are usually executed by subject matters experts (SME). An SME is an individual with a deep understanding of a particular job, process, department, function, technology, machine, material, or type of equipment. Once identified in the organization, these professionals can give support to keep the knowledge base updated and relevant.

The benefits of using social media to support knowledge management processes are disseminated in the literature, as presented in Fig. 3.2. Sarka and Ipsen (2017) affirmed that using SM to store and share knowledge can effectively help software developers to achieve project objectives. Nabelsi et al. (2017) reported project performance benefits from wiki usage in

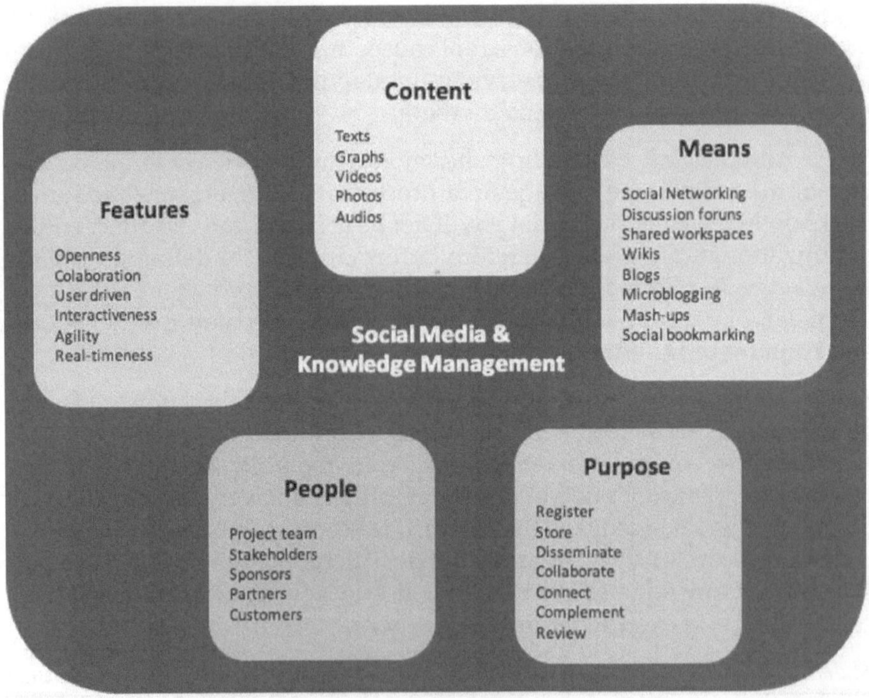

Fig. 3.2: Social media and knowledge management.
(Alt Text: A figure showing how social media through its content, features, means, purpose and people can be used to support knowledge management)

knowledge sharing within IT projects in the public sector. Foote and Halawi (2018) pointed out the different SM tools that aided the team members in developing higher-quality software. Chowdhury and Lamacchia (2019) presented a collaborative framework where social media tools facilitate sharing knowledge, contributing to successful digital transformation projects. Gomes et al. (2018) identified the technologies used for knowledge sharing (KS) within the project management process groups during the project life cycle. According to their findings, instant messages, Microsoft Office, PM tools, e-mail, Sharepoint, and teleconferencing tend to be used for KS throughout most PM process groups. Duffield and Whitty (2015) and Veronese and Chaves (2016) discuss the use of wikis to support the dissemination and application of lessons learned from projects.

The usage of social media improves the tacit and explicit knowledge of the decentralized team, thus increasing the team's connectivity (Margaryan et al., 2015). The transformation of tacit knowledge into explicit knowledge by transcribing knowledge contributes with benefits, transmitting non-visible and explainable knowledge (Takeuchi and Nonaka, 2004).

Social media could not only empower project organizations to collect progressively, register, store and share knowledge (such as that based on retrospection meetings and other experiences of project staff) through the active involvement of all project team members but can also be used to send out lessons learned to the relevant stakeholders on time (Narazaki et al., 2020). With the extensive adoption of social media in the workplace, supported by portable devices, historical project information is more easily accessible. It should be explored in order to reduce risks and enhance the quality of project work.

3.3 Method

We adopt a positivist stance using inductive reasoning, exploratory analysis, and qualitative techniques (Patton, 2015) to understand better the phenomenon of how senior project managers manage communication and knowledge sharing in remote IT projects. Table 3.1 details the main methodological aspects of this research.

The research sought to identify the practical impacts caused by the COVID-19 pandemic on projects, considering the use of social media for communication and knowledge sharing. Thus, with its roots in empiricism, positivism proposes aligning experience with human knowledge, where invariable laws govern social phenomena. In this sense, the interviewees describe the perception of the use of social media in the home/office work period.

Based on empirical investigation, the researchers encouraged IT professionals to provide data from their projects to help understand the issues of this research. We collected data by an exploratory approach

Table 3.1: Methodological Aspects of this Research.

Aspect	**In This Research**
Paradigm	Positivist
Research nature	Qualitative
Scientific approach	Inductive
Methodological approach	Exploratory
Unit of analysis	Use of social media for communication and knowledge sharing along with remote IT projects
Unit of observation	Senior project workers (several roles) working in companies from different sectors and countries (Mexico, Brazil, Germany, and Romania)
Temporal clipping	Cross-sectional
Data collection technique	Interviews
Data analysis	Content analysis

through interviews throughout the COVID-19 pandemic. Data collection was done through direct interviews, applied to project members chosen for their profile, professional experience, and seniority, identifying data patterns, and justifying explanations and conclusions (Haig, 1995). The nine interviews took place in a structured way, with open questions (*see* Appendix 3.1) and were guided by the researcher. The research presents a holistic and practical observation of phenomena in IT projects experienced during their tasks and activities.

We adopted inductive logic in data analysis, looking at all the answers given to each question. Considering this exploratory approach, we also opted to bring to the manuscript a comprehensive set of interview excerpts to allow further researchers to use this rich empirical data.

3.4 Data Analysis

The senior PM practitioners interviewed are all 38-years old, have on average 5.5 years working in the current organization, with 6.5 years of experience in the current role. The distribution of respondents by project area is 40% in software development; 40% in software development and infrastructure; 30% in development, infrastructure, service desk, field service; 10% in infrastructure and development projects; 10% in infrastructure projects. As for approaches, the distribution is: 30% work in hybrid and agile projects; 30% work on hybrid projects; 20% work only on a traditional project (waterfall); 10% work only on agile projects. This data analysis presents an overview of the use of social media by senior PM practitioners and how they use these tools for communication and knowledge sharing in projects.

Among the findings, it was evident in the descriptions, when exposing their experiences in the period of the COVID-19 pandemic, that the use of social media is being transformed and is replacing interaction processes, means of communication, and especially of control. Featured in all respondents who used or had started to use it during the COVID-19 pandemic period, Microsoft Teams as social media was indispensable to home/office work. The tools and the percentage that the interviewees identified are given in Table 3.2.

3.4.1 Overview on the Use of Social Media

All the respondents agree that the use of social media was significantly intensified throughout the COVID-19 pandemic. Table 3.2 presents the primary social media used by the interviewees.

Among the social media used by respondents, those used most intensively during the COVID-19 pandemic were Microsoft Teams and

Table 3.2: The Primary Social Media Used by the Interviewees.

Tool	Percentage of Usage
WhatsApp	70
Microsoft Teams	70
E-mail	30
SharePoint	30
Zoom	30
Google Meet	10
Wiki	50
Slack	20

WhatsApp, appearing in 70% of the reports. Teams was highlighted by three interviewees for having extensive use in several areas of knowledge. Similarly, but to a lesser extent, contributing mainly to passive communication, the use of wiki was increased during the COVID-19 pandemic, representing 50% of the mentions. At a lower level, 30% of respondents used e-mail, SharePoint and Zoom.

Interviewee 6 (I6) gives an example of the lives promoted by the company, "Until recently, talking in a meeting via Zoom was a complex thing; nowadays everyone knows about Zoom. We used to do two lives a year, now it's two a week." I9 adds that the use grows exponentially and "... the pandemic only came to highlight the traditional activities that no longer made sense, one of which was the way we communicate. For example, the majority of our professionals are millennials, and they are more accustomed to entering a social network than to following an e-mail."

On the benefits that communication in social media brings to project management, all the interviewees report how this is happening. I5 and I8 emphasize that social media brings together people who are working remotely. '...remotely bring people closer through chat, meetings, and contacts in general'. (I5) and '... the clients felt closer and more involved in the project...' (I8). This environment allows 'gains in time, scope, cost, planning, communication and integration in general.' (I5) and 'have an agile approach where continuous discussions on the issues allowed us to detect the problems in scope, time and cost in less time.' (I8). I6 also stresses time gained using social media: 'a question that you solve by following a guideline you found on a social network, in a blog, in a post, in a comment, saves you time. ... I would say that we should even use more.' I4 adds that social media '... contributes in scope, opens new horizons, enabling new projects'.

According to the interviewees, the major benefit generated by social media during the COVID-19 pandemic was communication. They mentioned that communication was agile, transparent, accessible, and dynamic. As social media allowed more access to people, it was possible to have more audience in the online meetings than in a physical one (I9). In addition to communication, I5 and I7 refer to the remote integration provided by social media. I7 adds that social media '… improved control, tracking of activities with Kanban.'

We also approached the interviewees regarding the support of top management and the environment in the organization to use social media in projects. They comment that top management stimulates the use, and I9 adds, 'The adoption policies are not very clear yet.' Finally, only two interviewees report restrictions to access SM, such as Facebook and Instagram regarding the companies' restrictions. All companies advise against disclosing confidential data and information, according to the interviewees.

3.4.2 The Use of Social Media in Communication

Project communication supported by social media is a research theme that needs to be better explored by scholars. We start by identifying which tool is used in which type of communication. Table 3.3 shows that MS Teams, WhatsApp, and Yammer are the more frequent tools used to communicate. As expected, informal communication is carried out by WhatsApp, while MS Teams and Yammer enable formal communication. MS Teams and WhatsApp support active communication, while passive communication is carried out with other tools, such as SharePoint, wiki, and Slack. It is noteworthy that respondents also mention many other tools.

In addition to identifying the social media used in the four types of communication, we also asked the interviewees about the role of these tools in PM. The results indicate that the use of social media to support active and passive communication is pervasive throughout the project life cycle. Regarding active communication, I3 reports using Teams to support executive meetings, I1 mentions that WhatsApp is used to send reports, and I6 asserts that social media 'are the base'. Interviewees using agile approaches, such as Scrum, use social media to support ceremonies, daily meetings, planning, and review. When asked about passive communication, MS Teams was the most frequent tool mentioned by the interviewees, followed by Yammer, wiki, SharePoint, and WhatsApp. These tools are used as knowledge bases and to document manuals, templates, procedures, and test results.

Table 3.3: Number of Mentions Crossing Types of Communication and Collaborative Tools.

Tool	Type of Communication				Total Number of Mentions
	Formal	Informal	Active	Passive	
WhatsApp	2	7	5	2	16
MS Teams	6	3	7	4	20
Yammer	3		2	2	7
Jira	1		1	1	3
SharePoint	1		1	2	4
Wiki	2			2	4
Slack	1		1	1	3
LinkedIn		1	1		2
E-mail	2				2
Chame.in	1				1
Kuip	1				1
Service Now	1				1
Chime			1		1
Miro			1		1
HP (Test Tool)				1	1
Octane (Test documentation tool)				1	1
Telegram				1	1

Interviewees suggest the best practices for communication:

- **Take into account the context in which the communication is happening:** I7 mentions it is essential to have a deep understanding of who the attendees will be, what the context is, what the main objectives are, and who will act as a moderator and facilitator. I8 adds, "In geographically dispersed teams, it is crucial to standardize the language, avoiding local idiomatic expressions to ensure that the entire team understands the information."
- **Time definition:** "Define the schedule in which communications should take place since we can no longer assume that our schedule is the same as of the others." (I8)
- **Exclusive use of tools for work subjects:** "Awareness of use restricted to working in MS teams, for example not talking about football, politics, etc…" (I5)

- **Transparency and clear processes:** I6 suggests, "The sooner I show you that there is a problem, the faster it will be resolved." I8 adds, "Do not ignore, always establish very clear processes, as well as ways to audit their proper operation."
- **Culture:** I8 mentions, "Knowing the culture of others, our culture is part of our identity and we have to know the culture of the people who are part of our work team." I6 adds, "When you have projects here in Brazil, being direct, often (hurts) offends people. It is something that does not create a problem when I am with an American project."
- **Sense of team:** "Generate a sense of team through dynamics that allow the team to recognize themselves as people. We generally become more empathetic when we realize that other people are just as human as we are." (I8). Self-organizing groups are more effective and efficient in finding solutions; therefore, it is important to have a facilitator but to give the team the freedom to define rules and the way that they will work. Communities are not imposed; they are created around common objectives and interests" (I7).
- **Status update by recurrence:** I6 stresses that we should "always have the discipline to have the status meeting with the staff." I3 and I4 reinforce the practice of daily meetings to make the status of the project clear. In this context, the team needs to define the tool that will be the single source of status information and keep it updated. I7 related that using multiple tools in a project made the follow-up activity complex since the information needed to be updated in many different repositories.

3.4.3 The Use of Social Media to Share Knowledge

In addition to communication, social media is used to share knowledge. Interviewees were unanimous in affirming that remote work cannot exist without collaborative tools and that social media can improve knowledge sharing. They mentioned the following aspects to explain how this can occur: training by using YouTube videos (I5, I6, I7, and I8), improving the quality of content to be published (I4), creating communities (I7), sharing knowledge in a structured way (I4), agility (I2 and I5), and facilitating the sharing of project documents and lessons learned (I3). I7 adds, "All tools become knowledge base" and concludes by saying that "social media are providing a revolution in knowledge sharing."

When asked about other social media tools to share knowledge, interviewees add Git, wiki, and Slack. Git is a 'software for tracking changes in any set of files, normally used for coordinating work among programmers collaboratively developing source code during software development'. The wiki should better disseminate knowledge and its evolution, change, and dynamism (I5). Slack is a collaborative tool designed for use in the workplace. It allows for direct messages between two individuals and group chats where many people can share content, comments, and reactions.

Although most of the respondents mention tools, I4 assert, "There is no lack of tools, but rather there is a need to improve the knowledge management processes inside organizations and mechanisms to motivate people to be knowledge generators. There is a lack of processes that formalize project records for the dissemination of knowledge." In this same line, I7 add, "There is no need for new tools, but you should have more freedom to use the tools. Tool integration is important. So as not to reinvent the wheel every time you need to use a new tool."

On the other hand, interviewees reported their main concerns about the use of social media to share knowledge as follows:

- *Knowledge management processes:* The interviewees consider the collaborative features of mentioned tools one of the essential advantages to sharing knowledge among teams. On the other hand, almost all interviewees affirm that there is no formal process in place. Moreover, they recognize that if the organization has a formal process, it will help them form new subject matter experts, reduce risks, and improve estimations.
- *Integration:* Some interviewees mention the importance of integrating tools to improve productivity and avoid inaccuracies (different versions in different repositories), rework, and excessive notifications.
- *Information security and confidentiality:* I2 reports his main concern is information leakage (I2), while I4 asks for "ontrol and rules to define what can be published. Confidential information must be preserved." I6 draws our attention to prevent the leakage of strategic information.
- *Quality of information:* The main aspects of the quality of information mentioned by the interviewees are veracity and reliability. I9 mentions, "One of the greatest challenges that social networks have is verifying the truthfulness of the data that is disclosed; it is a huge commitment that administrators must adopt to prevent misleading the public or promoting crimes or false accusations. In the case of knowledge exchange, the information found must come from reliable sources." The quality of information shared is also a concern for I1, who reinforces the importance of subject matter experts or other curators as knowledge managers to review the information before being distributed, assuring the accuracy and relevance of the information.
- *Notification overload generating stress:* Notification overload occurs due to the use of multiple tools which generate "many notifications related to the same project," as reported by I7. I8 complements that

sometimes "this information overload does not generate value and can impact team's productivity."

- *Transparency:* Transparency is often related to communication, but this time was mentioned as a concern related to knowledge sharing. Here transparency appears as a requirement for the organizations to establish and communicate transparent knowledge management processes. Moreover, interviewees also mentioned the importance of clear rules to adopt new collaborative tools, thereby avoiding risks of shadow IT.

We also asked the interviewees about the existing agreements to use social media to share knowledge. I3 and I5 suggested maintaining only the information related to the projects in the social media. "Define the scope and subjects that can be shared in social media to keep relevant information," said I3. "Define with the team the best practice to share information, maintain courtesy and professional language. Do not bring information that is out of the scope," advises I5.

I1 and I7 point out that there is a need to organize information to facilitate indexing and searching. I1 uses the WhatsApp tool as an example: the information shared in this social media must be organized to be preserved and easily accessed.

I9 suggests the definition of policies as knowledge evaluation mechanisms with curators that ensure the integrity and veracity of the information, such as: 1. responsible use of the tool; 2. disclosure standards; 3. material prohibited or not pleasant to the community; 4. documentation not reviewed; and 5. documents from unreliable sources.

Finally, we asked the interviewees about the initiatives to convert tacit knowledge to explicit knowledge. They mention the importance of initiatives that motivate employees to share their experiences and how social media can play an essential role in identifying and forming new subject matter experts. I1 affirms that recognizing the top, more visible contributors in the organizational environment, is one of the best approaches to motivating people to create relevant content. I3, I5, and I7 discuss how the collaborative features of the social media tools provide content feedback, making employees produce content together, and create a positive cycle to feed the network continuously. I6 mentions how social media is being applied in his company in the transfer knowledge method, ensuring that everyone in operation has the information they need to run smoothly. "There is a process that is the KT (knowledge transfer). A person who is going to leave the company has to document everything he did for a given client. Then another person comes in, and he can understand all the actions reported in the KT," says I6.

3.5 Discussion

The analysis of the results generated some issues that deserve better attention by the PM community. We identify four issues to focus on the discussion: first, the need for investment by the organization for formalizing communication, mediation, and promotion of lessons learned; second, integration among the collaborative tools; third, the creation of formal procedures to register, store, share, and review knowledge; and fourth, identification of curators (subject matter experts) for reviewing the information and ensuring quality and trustworthiness.

Results already observed in an empirical study before the pandemic by Winter and Chaves (2017) registered an increase in the productivity of the project teams using social media. Not neglecting social media, in essence, brings informality as a basis, governed by the social behavior and personality of each group (Ikemoto et al., 2020). Based on the interviewees' reports, the importance of communication between project members is widely observed (Taibi et al., 2018), amplified by the effect of the COVID-19 pandemic on home/office working, supported by social media. However, interviewees emphasize that an organization's investment is necessary for formalizing communication, mediation, and promotion of lessons learned.

Although the various social media available are sufficient for the needs of the project teams, respondents report the lack of integration among the collaborative media on projects. Adding to the gap found in the lack of social media integration, this finding suggests that an organization-oriented communication model can improve communication in projects through social media. This model should address the need for storing knowledge and lessons learned, active collaboration, and formal and informal communication (Nothhaft and Schölzel, 2015). In the model, the role of the mediator should be considered in several stages (Cheung et al., 2013). Ikemoto et al. (2020) approach the integration of social media in projects. Most of the interviewees in our study report teams as a tool to affect this integration.

Successful organizations recognize the power of an effective knowledge management process to maintain their competitive advantage (PMI, 2015). Our interviewees emphasize that social media can revolutionize the way we share knowledge nowadays. However, it seems that there is still a lot to be done in this field. Observing the interviewees' answers, it is possible to perceive that teams working in projects related to the service desk and field services have the culture and discipline to share knowledge since they usually apply frameworks as Information Technology Infrastructure Library (ITIL) that have very well-defined knowledge management processes. On the other hand, experienced project managers working with application development, applying modern Agile frameworks, reported struggling to motivate their team to share project knowledge periodically. Agile frameworks usually define ceremonies as retrospective meetings whose objective is to identify with the team the main lessons learned. However, this kind of framework does not define knowledge management processes—resulting in a lack of formal procedures to register, store, share, and review knowledge.

Organizations that are effectively exploiting social media for knowledge sharing have a culture that values such a practice. Leaders support and encourage the culture of creating, sharing, usage, promotion, and acquiring new knowledge in organizations. Interviewees were unanimous about the importance of defining mechanisms to ensure information security and confidentiality. IT managers need to establish what tools can be used and how they can be integrated, avoiding potential security threats to the organization. Moreover, interviewees agree about the importance of one or more curators (subject matter experts) reviewing the information and ensuring quality and trustworthiness. Another challenge, pointed out by our interviewees, is how to prepare the new generation to be the next subject matter experts and how social media can be applied, improving the employees' visibility and consequently motivating them to share knowledge. The study published by the PMI research group (PMI, 2015) mentions the importance of building a multigenerational workplace and how crucial knowledge sharing is. As the population ages and millions begin to retire, companies stand to lose a wealth of accumulated knowledge.

3.6 Conclusion

While there is a broad field of knowledge management research and many studies presenting the benefits of social media in the project management area, not much has been published about the challenges of managing knowledge by using social media. This research begins to fill that gap in understanding how to fully explore the potential of social media to support communication and knowledge sharing during remote work.

This study is not free of limitations, which include the small number of interviews and the remote environment in which they were carried out. The analysis of the results opens new issues for research, such as the use of integrated collaborative tools, policies about information security and confidentiality, and the need to foster the process of formalizing the communication, mediation, and promotion of lessons learned.

References

Ahmed, Y.A., Ahmad, M.N., Ahmad, N., and Zakaria, N.H. (2019). Social media for knowledge-sharing: A systematic literature review, *Telematics and Informatics*, **37**(January 2018): 72–112.

Anwar, R., Rehman, M., Wang, K.S., and Hashmani, M.A. (2019). Systematic literature review of knowledge sharing barriers and facilitators in global software development organizations using concept maps, *IEEE Access*, **7**(c): 24231–24247.

Anwer, F., Aftab, S., Shah Muhammad Shah, S., Waheed, U., Shah, S.M., and Waheed, U. (2017). Comparative analysis of two popular agile process models: Extreme programming and Scrum, *International Journal of Computer Science and Telecommunications*, **8**(2): 1–7. Retrieved from https://www.ijcst.org/Volume8/Issue2/p1_8_2.pdf.

Banjo, S., Yap, L., Murphy, C., and Chan, V. (2020). The Coronavirus Outbreak has become the World's Largest Work-from-Home Experiment. Available at: <https://time.com/5776660/coronavirus-work-from-home/?utm_source=link_wwwv9&utm_campaign=item_288956&utm_medium=copy>. [accessed on 17 November 2020].

Chaves, M.S., Scornavacca, E., and Fowler, D. (2018). Affordances of social media in knowledge sharing in intra-organisational information technology projects. pp. 35–47. *In: Digital Technology and Organizational Change*, Springer.

Cheung, S.O., Yiu, T.W., and Lam, M.C. (2013). Interweaving trust and communication with project performance, *Journal of Construction Engineering and Management*, **139**(8): 941–950. https://doi.org/10.1061/(asce)co.1943-7862.0000681.

Chowdhury, K., and Lamacchia, D. (2019). *Collaborative Workspace for Employee Engagement Leveraging Social Media Architecture*, Society of Petroleum Engineers—Abu Dhabi International Petroleum Exhibition and Conference 2019, ADIP 2019.

Duffield, S., and Whitty, S.J. (2015). Developing a systemic lessons learned knowledge model for organizational learning through projects, *International Journal of Project Management*, **33**(2): 311–324.

Dwivedi, Y.K., Wastell, D., Laumer, S., Henriksen, H.Z., Myers, M.D., Bunker, D., ... and Srivastava, S.C. (2015). Research on information systems failures and successes: Status update and future directions, *Information Systems Frontiers*, **17**(1): 143–157.

Foote, A., and Halawi, L.A. (2018). Knowledge management models within information technology projects, *Journal of Computer Information Systems*, **58**(1): 89–97.

Gholami, B., and Murugesan, S. (2011). Global IT project management using Web 2.0, *International Journal of Information Technology Project Management*, **2**(3): 30–52. doi: 10.4018/jitpm.2011070103.

Gomes, F., Oliveira, M., and Chaves, M.S. (2018). An analysis of the relationship between knowledge sharing and the project management process groups, *Knowledge and Process Management*, **25**(3): 168–179.

Hacker, J.V., Johnson, M., Saunders, C., and Thayer, A.L. (2019). Trust in virtual teams: A multidisciplinary review and integration, *Australasian Journal of Information Systems*, 23.

Haig, B.D. (1995). Grounded theory as scientific method, *Philosophy of Education*, 28(1): 1–11.

Henderson, L.S., Stackman, R.W., and Lindekilde, R. (2016). 'The centrality of communication norm alignment, role clarity, and trust in global project teams', *International Journal of Project Management*, **34**(8): 1717–1730.

Ikemoto, M.N., Gantman, S., and Chaves, M.S. (2017). Use of social media in IT project management: A literature review based on hermeneutics and a research agenda, *Iberoamerican Journal of Project Management*, **8**(1): 87–107.

Ikemoto, M.N., Chaves, M.S., Gantman, S., and Russo, R.D.F.S.M. (2020). SM4PM: A prescriptive framework for guiding integrated use of social media in project management, *International Journal of Management and Enterprise Development*, **19**(4): 330–364.

Iriarte, C., and Bayona, S. (2020). IT projects success factors: a literature review, *International Journal of Information Systems and Project Management*, 8(2): 49–78.

Kanagarajoo, M.V., Fulford, R., and Standing, C. (2019). The contribution of social media to project management, *International Journal of Productivity and Performance Management*, 69(4): 834–872.

Kane, G.C. (2017). The evolutionary implications of social media for organisational knowledge management, *Information and Organisation*, **27**(1): 37–46.

Krumova, M.Y., and Milanezi, B.D. (2015). Knowledge Sharing & Collaboration 2.0, *KSI Transactions on Knowledge Society*, **7**(4): 51–56.

Lanubile, F., Ebert, C., Prikladnicki, R., and Vizcaíno, A. (2010). Collaborative tools for global software engineering, *Software IEEE*, **27**(2): 52–55.

Majdenbaum, A., and Chaves, M. (2020). Social interaction promotion in distributed software development in agile projects. *In: Proceedings of the 28th European Conference on Information Systems (ECIS)*, an online AIS conference, June 15–17.

Margaryan, A., Boursinou, E., Lukic, D., and de Zwart, H. (2015). Narrating your work: An approach to supporting knowledge sharing in virtual teams, *Knowledge Management Research & Practice*, **13**(4): 391–400.

Maxwell, J.A. (2008). Designing a qualitative study, *The SAGE Handbook of Applied Social Research Methods*, 2: 214–253.

Nabelsi, V., Gagnon, S., and Brochot, D. (2017). Increasing the impact of wikis on project performance: Fine-tuning functional quality and knowledge sharing, *Knowledge and Process Management*, **24**(2): 96–113.

Nach, H. (2016). Project management 2.0: Towards the renewal of the discipline. pp. 1–15. *In:* Silvius, G. (Ed.). *Strategic Integration of Social Media into Project Management Practice*, Hersey, PA: IGI Global.

Naeem, M. (2019). Uncovering the role of social media and cross-platform applications as tools for knowledge sharing, *VINE Journal of Information and Knowledge Management Systems*, **49**(3): 257–276. https://doi.org/10.1108/VJIKMS-01-2019-0001.

Narazaki, R.S., Chaves, M.S., and Pedron, C.D. (2020). Social media in knowledge management: A holistic knowledge funnel based on a retrospective 10-year study in top-tier journals, *Knowledge and Process Management*, **27**(2): 123–132.

Nesheim, T., and Hunskaar, H.M. (2015). When employees and external consultants work together on projects: Challenges of knowledge sharing, *International Journal of Project Management*, **33**(7): 1417–1424.

Nothhaft, H., and Schölzel, H. (2015). (Re-) Reading clausewitz: The strategy discourse and its implications for strategic communication, *The Routledge Handbook of Strategic Communication*, Routledge, January 2015, 18–33.

Nyström, C.A., and Asproth, V. (2013). Virtual teams – support for technical communication? *Journal of Organisational Transformation & Social Change*, **10**(1): 64–80.

Ozguler, I.S. (2020). Insights report - project management after COVID-19, *PM World Journal*, IX(V): 1–20.

Park, J.G., and Lee, J. (2014). Knowledge sharing in information systems development projects: Explicating the role of dependence and trust, *International Journal of Project Management*, **32**(1): 153–165.

Patton, M.Q. (2015). *Qualitative Research and Evaluation Methods*, Sage, 4th ed., 806 pages.

PMI. (2013). *Pulse of the Profession In-depth Report: The Essential Role of Communications*, Project Management Institute.

PMI. (2017). *A Guide to the Project Management Body of Knowledge (PMBOK® Guide)*, 6th ed., PMI.

PMI. (2020). *Pulse of the Profession In-depth Report: Ahead of the Curve: Forging a Future Focused Culture*, Project Management Institute.

PMI. (2021). *Pulse of the Profession In-depth Report: Beyond Agility: Flex to the Future*, Project Management Institute.

Price, J.L. (1997). Handbook of organizational measurement, *International Journal of Manpower*, **18**(4/5/6): 305–558.

Qureshi, R., Basheri, M., and Alzahrani, A.A. (2018). Novel framework to improve communication and coordination among distributed agile teams, *International Journal of Information Engineering and Electronic Business*, **10**(4): 16.

Rimkuniene, D., and Zinkeviciute, V. (2014). Social media in communication of temporary organizations: Role, needs, strategic perspective, *Journal of Business Economics and Management*, **15**(5): 899–914.

Rosa, D.V., Chaves, M.S., Oliveira, M., and Pedron, C. (2016). Target: A collaborative model based on social media to support the management of lessons learned in projects, *International Journal of Managing Projects in Business*, **9**(3): 654–681.

Sarka, P., and Ipsen, C. (2017). Knowledge sharing via social media in software development: A systematic literature review, *Knowledge Management Research and Practice*, **15**(4): 594–609. https://doi.org/10.1057/s41275-017-0075-5

Sun, Y., Zhou, X., Jeyaraj, A., Shang, R.A., and Hu, F. (2019). The impact of enterprise social media platforms on knowledge sharing: An affordance lens perspective, *Journal of Enterprise Information Management*, **32**(2): 233–250.

Taibi, D., Lenarduzzi, V., Ahmad, M.O., and Liukkunen, K. (2017). Comparing communication effort within the Scrum, Scrum with Kanban, XP, and Banana development processes, *ACM International Conference Proceeding Series, Part F128635* (August), 258–263. https://doi.org/10.1145/3084226.3084270

Takeuchi, H., and Nonaka, I. (2004). Knowledge creation and dialectics, *Hitotsubashi on Knowledge Management*, 1: 1–27.

Valladares, L. (2007). *Os dez mandamentos da observação participante, Revista brasileira de ciências sociais*, **22**(63): 153–155.

Van Ruler, B. (2021). Communication planning: Agility is a game changer in strategy development, *International Journal of Strategic Communication*, **15**(2): 113–125. https://doi.org/10.1080/1553118x.2021.1898117

Veronese, G.T., and Chaves, M. (2016). An integrated approach to collaborative learning in projects. *In Strategic Integration of Social Media into Project Management Practice* (pp. 160–178). IGI Global.

Winter, R., and Chaves, M.S. (2017). Innovation in the management of lessons learned in an IT project with the adoption of social media, *International Journal of Innovation: IJI Journal*, **5**(2): 156–170.

Yin, R.K. (2011). *Qualitative Research from Start to Finish*, Guilford Press.

Zhang, Y., Sun, J., Yang, Z., and Wang, Y. (2018). Mobile social media in inter-organizational projects: Aligning tool, task and team for virtual collaboration effectiveness, *International Journal of Project Management*, **36**(8): 1096–1108.

Appendix 3.1—Interview Protocol

Acknowledgement (the interview will be recorded and anonymous)

1. Name
2. Age
3. Organization
4. The number of years in this organization
5. Role
6. The number of years in this role

7. Types of projects you have been working on: Development, Infrastructure, Service Desk, Field Service…)
8. Approaches used (agile, traditional, hybrid)
9. What do you understand about social media?
10. Explain what kind of social media we are considering.

All your answers should be given considering a project being executed remotely. The theoretical grounding of each interview question is summarized in Table 3.4.

Table 3.4: Theoretical Grounding of Interview Questions.

	Question	**References**
Social Media	1—What are the primary social media used in the projects in which you participate or participated?	Gholami and Murugesan (2011)
	2—Are the most used social media formally recognized by the company as a tool? Does your company restrict the use of any social media?	Duffield and Whitty (2015)
	3—Do projects benefit from communication through social media (deadline, scope, cost)?	Delerue and Sicotte (2017)
	4—Does the organization's leadership and environment contribute to social media use?	Delerue and Sicotte (2017)
	5—Was the use of these social media intensified during the COVID-19 pandemic?	
	6—Cite the main benefits of the use of social media during the COVID-19 pandemic.	
Communication	7—What social media does your team use for formal/informal communication?	Lanubile et al. (2010); Zhang et al. (2018)
	8—Explain the role of social media in the direct communication of your team (daily meeting, status meeting, review and retrospective, lessons learned), for example.	Anwer et al. (2017); Taibi et al. (2018)
	9—What social media does your team use for direct communication?	Anwer et al. (2017); Taibi et al. (2018)
	10—Explain the role of social media in your team's passive communication (training material, source code documents, test documents).	
	11—What social media does your team use for passive communication?	
	12—What are your best practices for communication to flow smoothly throughout the project being executed remotely?	

Table 3.4 contd. ...

...Table 3.4 contd.

	Question	**References**
Knowledge Sharing	13—Do you think social media can be used to improve knowledge sharing in IT projects? Why? Why not?	
	14—In your opinion, besides the tools already mentioned, what other social media should be applied to improve knowledge sharing?	
	15—What are your main concerns about applying this kind of tool to knowledge sharing?	
	16—What kind of agreements or definitions do you think are essential to have the best usage of social media in knowledge sharing?	
	17—How do you think organizations can motivate their employees to use social media to transform tacit into explicit knowledge?	

4

Social Media for Project Knowledge Management

Jianyao Jia, Guofeng Ma* and *Shan Jiang*

4.1 Introduction

Social media refer to a group of internet-based technologies that allow users to easily create, edit, evaluate, and/or link to content or other creators of content (Kaplan and Haenlein, 2010). In practice, social media have penetrated into people's daily life as well as workplace (Brooks and Califf, 2017; Ahmed et al., 2019). Undoubtedly, social media have also been adopted in project management practice for easy access and low cost (Pivec and Maček, 2019). Previous research has argued that social media have the potential to change traditional knowledge-management paradigm, mostly due to the characteristic of communication visibility afforded by social media (Leonardi, 2014). In this way, social media could facilitate project knowledge management, which has been verified by empirical research in the project management field (Jia et al., 2020; Ma et al., 2021).

This chapter will summarize the current work concerning social media for project knowledge management. Specifically, the next section focuses on how project managers (individual) could employ social media for knowledge acquisition. The third section concentrates on how project teams use social media for team knowledge sharing. After that, how the

Tongji University, China.
* Corresponding author: jianyaoJ@tongji.edu.cn

project could utilize social media to realize knowledge integration is discussed. This chapter ends with the conclusion and future research.

4.2 Social Media for Project Managers' Knowledge Acquisition

Project work is challenging due to the project nature (i.e., uniqueness, temporality, one-time). In other words, the project itself is set up to deal with uncertainties and risks. From the knowledge-based perspective, project management is the application of knowledge to project activities to meet the project requirements (PMI, 2017). In practice, even experienced project managers can't hold or master all required knowledge for project tasks. In this way, it is necessary for project managers to acquire additional knowledge to complete work tasks (Yu et al., 2018).

Previous research had classified social media into work-oriented social media (e.g., DingTalk) and socialization-oriented social media (e.g., WeChat) according to the differences in technical features, affordances, and supported ties (Song et al., 2019), both of which can be used by project managers for work purposes and knowledge acquisition. Project managers can easily access these mobile applications, which can be employed by project managers to promote their knowledge acquisition. Specifically, in practice, it is common for project managers to encounter similar problems which have appeared in previous projects (Oti et al., 2018). Yet, these problems are commonly solved in a team-based way. Put it another way, the knowledge needed to solve these problems resides in previous teammates' experiences. Since the project manager has only part knowledge on these problems, it is almost impossible for her or him to solve them independently. Luckily, the project manager has the knowledge of who can help to solve the particular problem (i.e., the knowledge of who knows what). Thus, the project manager may need to acquire knowledge from former participants to solve the problem. In this situation, work-oriented social media can support knowledge acquisition from former workmates in the same company (e.g., enterprise social media), and socialization-oriented social media can support knowledge acquisition from employees of other companies (e.g., WeChat). As prior research indicates, social media use in the project context helps establish social ties (Huang et al., 2020), by which project managers could contact former teammates from their joint participated projects. In practice, both work-oriented social media and socialization-oriented social media facilitate knowledge acquisition as they enable watching online videos, live chat, sending instant messages, etc. (Ahmed et al., 2019; Rambe and Nel, 2015). These features have great potential to enable both explicit and tacit knowledge acquisition. For instance, photo presentation helps clarify the encountered problems in a visualized way. The underlying logic lies that visualization decreases communication barriers (e.g., geographical barriers and knowledge diversity) by enabling a common language between them. The characteristic of live chat helps remove the limits of time and space, which are commonly viewed as barriers in knowledge exchange. On the other hand, project managers can have access to external knowledge by searching in wikis and YouTube, etc. In today's sharing economy era, it is easy for us to acquire knowledge by social media. Online sharing platforms, such as wikis, Zhihu, and professional online communities provide excellent opportunities to access knowledge at low cost. Although the problems that project managers encounter may be unique, previous experience and accumulated knowledge are still informative in problem solving. Of note, such public social media platforms are especially useful for explicit knowledge acquisition, including government regulations and industry norms.

Apart from external knowledge, social media also function as an effective tool for project managers to acquire the required internal knowledge to perform a task. Leonardi (2014) illustrated that work-oriented social media within an organization provides a forum for public communication among employees, thereby enabling communication visibility, which can increase the accuracy of people's metaknowledge. He qualitatively identified two important interrelated mechanisms in such a process: message transparency and network translucence. After that, the same author validated these findings with a quasi-natural field experiment (Leonardi, 2015). These studies provide strong theoretical underpins for knowledge acquisition in the project context. A recent survey has also indicated that building an online community is a significant purpose of social media use in projects (Pivec and Maček, 2019). In this way, project managers can contact the one from whom they want to acquire knowledge easily. Besides, informal online groups supported by socialization-oriented social media also promote internal knowledge acquisition. Project members feel free to talk in these informal social media groups, though there may exist some differences between formal and informal social media groups, such as shared knowledge and members within the groups, both of which could help acquire knowledge. To be specific, in formal social media groups, project managers tend to acquire explicit knowledge for its communicative feature. This is also to decrease information overload in formal social media groups for project managers. Compared to formal social media groups, informal social media groups are more suitable for tacit knowledge acquisition. Informal social media groups can be flexibly established by the project manager. Members of such groups are dependent on knowledge demands of the project manager. In this way, members of these informal social media groups could jointly help the project manager acquire the needed knowledge. Even in non-work social media communities, the triggered-attending affordance of social media (Majchrzak et al., 2013) enables those interested in a specific topic to engage in the online knowledge conversation, especially when project managers encounter tough issues. In such conversations, more or less, project managers will also acquire internal knowledge.

4.3 Social Media for Project Teams' Knowledge Sharing

As stated above, project work is team-based. Knowledge residing within each project team member could be homogeneous or complementary. In any case, sharing individual acquired knowledge with teammates is beneficial to team functioning. On the one hand, knowledge sharing enriches the knowledge base of the project team, thus potentially improving the team ability. On the other hand, knowledge sharing makes teammates aware of one's knowledge structure and transcend knowledge differences among team members, thus forming the foundation of common knowledge. A common practice in knowledge sharing is to employ innovative information technologies (Chang et al., 2017; Joseph Garcia and Mollaoglu, 2020). Thus, social media enables knowledge sharing within project teams.

Knowledge sharing is defined as the communication of knowledge from a source in such a way that it is learned and applied by the recipient (Ko et al., 2005). A survey conducted by Azhar and Abeln (2014) indicates that social media tools are being used to improve communication in the construction industry as social media tools enable instant feedback and two-way communication. Hasan et al. (2019) argued that social media technologies have changed how knowledge sharing takes place in projects due to their mobility. Leonardi (2017) suggested that social media facilitate knowledge sharing because they are leaky pipes for communication. Such a metaphor depicts how knowledge sharing happens in practice, which may also explain the reason why social media facilitate tacit knowledge sharing. Tacit knowledge is also known as the knowledge with regard to 'know-how'. In other words, tacit knowledge means the knowledge of how to perform the specific task. In social media, project managers could have the full landscape on how the task proceeds until its completion. During the course of problem solving, project managers themselves may also join the process to share their knowledge. In this way, following team knowledge sharing, knowledge application, knowledge creation, and other knowledge activities occur within the project team.

From another perspective, trust is the premise of knowledge sharing (Reagans and McEvily, 2003; Chiregi and Navimipour, 2016). Previous research has indicated that when people cannot see what others do, or when, with whom, and how they do it (i.e., visibility), it could be harder to form interpersonal trust among coworkers (Cramton et al., 2007). As suggested above, the main feature of social media for knowledge management lies in communication visibility. In this vein, engagement in team social media improves interpersonal trust and cultivates a trust atmosphere. Moreover, informal social interactions are also crucial in the development of interpersonal trust (McAllister, 1995). The bottom-up characteristic of social media adoption paves the way for informal interactions. In the literature, scholars have acknowledged the power of informal information within organizations and therefore explored this issue from the perspective of informal interaction (Neeley and Leonardi, 2018). Their analysis indicated that employees' curiosity about interactions motivates them to use social media. Then, the integration of non-work and work content allows employees to acquire their required knowledge and gauge passable trust. It is the passable trust that enables them to share knowledge. Compared to traditional information technologies, such as knowledge management systems, social media enable both synchronous and asynchronous knowledge sharing, while traditional informational technologies facilitate asynchronous knowledge sharing. In fact, project work is challenging and time limited as spending a lot of time on searching knowledge in knowledge management systems is costly for project managers. Further, although project managers could access related knowledge, it is difficult for them to apply such knowledge in their tasks directly. A recent study also indicated that socialization-oriented social media facilitates team knowledge sharing (Song et al., 2019). Although members in projects are likely to be unfamiliar with each other, social media-enabled visibility and informal interactions will improve trust among project teams to promote knowledge sharing. Knowledge sharing by social media overcomes shortcomings in traditional knowledge management systems, which is also the reason that social media-enabled knowledge sharing has received great attention from both academic and practice.

Prior scholarship has suggested that physical co-location of team members plays a key role in fostering knowledge sharing and improving team performance. However, physical co-location poses great difficulties for many teams and the cost of such work way is great, especially for some distributed project teams. Social media provides an opportunity for team members to co-locate in a virtual space. With the same logic, the proximity to other team members makes them have a sense of psychological safety and a feeling of belongingness. In such circumstances, knowledge sharing occurs naturally. Of note, it is also observed that team members are silent in many online communities supported by social media. In other words, knowledge sharing within team social media is also dependent on other factors. Social media, as a set of tools, will not improve knowledge sharing automatically. A possible effective way to make knowledge sharing work in social media platforms is to employ hidden knowledge facilitators (Li-Ying et al., 2018). The underlying theoretical underpin for this measure lies in the domain of organizational culture. The knowledge sharing behavior of hidden knowledge facilitators provides an additional clue for other team members. As long as some other online team members react to a hidden knowledge facilitator's post, a bandwagon effect can potentially emerge (Sundar, 2008).

4.4 Social Media for Project Knowledge Integration

The knowledge-based view suggests that knowledge integration is the fundamental capability of an organization (Grant, 1996). Organizations have to integrate specialized knowledge to realize knowledge application, which embodies in products or services. Similarly, performing a project commonly involves various types of knowledge. The project will have the potential to be successful only when these knowledges are integrated and applied to project tasks. In this way, knowledge integration plays a central role in performing projects (Jia et al., 2021).

Knowledge integration is defined as the process of synthesizing different types of (specialized) knowledge (Okhuysen and Eisenhardt, 2002). On the one hand, social media function as a knowledge pool into which project participants can put their knowledge. This forms the foundation of project knowledge integration. In practice, with the increase in the size of social media online communities (i.e., the number of members in such communities), the efficiency of project knowledge management by such tools tends to decrease. The information processing theory provides an explanation for this phenomenon. Different from large-scale open online communities, project online communities are established to fulfil project requirements, which have specific objectives and aims to solve project issues. In this way, it could be appropriate to host a project online community which involves boundary spanners from each party. The boundary spanner acquires knowledge, not only for herself or himself, more importantly, for knowledge sharing within the project team and knowledge integration within the project online community. The boundary spanner occupies the role of a broker who links the local team and the whole project (Pemsel and Wiewiora, 2013). Or in other words, the boundary spanner promotes knowledge flows across different levels (individual, team, and project).

On the other hand, from the micro-perspective, knowledge integration could occur in project online communities in two ways: proactive and reactive. To be specific, although project parties occupy their own tasks, but their tasks are commonly mutually dependent. In this way, when one project party encounters problems, it is necessary to consider how to coordinate with other parties. Such a process requires knowledge integration among the parties. The party could present the problem in a visualized way (i.e., photo or video) in the project online community. Then, the party may contact relevant parties to integrate their knowledge to jointly solve the problems at hand. In a similar way, the project party could be required to contribute their knowledge to solve other parties' problems. In this situation, the project party will achieve knowledge integration in a reactive way.

Up to now, this chapter has illustrated how social media could be utilized for knowledge management in the context of project, including individual-level, team-level, and project-level usage. It should be noted that social media use at different levels is correlated. The possible relations are summarized as follows. In short, social media are used by project managers to acquire knowledge. Then, both the acquired knowledge and previously owned knowledge by the project manager could be shared within teams by social media tools. At the project level, social media are utilized to integrate knowledge from different parties. The embedded relationships provide rationale for social media use in knowledge management across different levels in the project context.

4.5 Conclusion and Future Research

While previous research has explored a lot on social media for knowledge management, social media for project knowledge management is still an emerging field deserving further investigation. Although several studies have affirmed the benefits of social media use for project knowledge management, the micro dynamics are missing. Put it another way, little is still known about what happens in project knowledge management practice. Besides, the cross-level knowledge flows among projects and across projects by social media require further investigation. Due to the bottom-up characteristic of social media use, managers may not be able to figure out what really happens in practice. In this way, it is necessary to explore how micro dynamics evolve into higher-level phenomenon. For instance, project members are active knowledge contributors in some online project communities, while they are silent in other communities. What factors contribute to such behavior differences in the project context is also a research question deserving investigation.

Previous scholarship has put great efforts to explore knowledge-sharing behavior on social media platforms, while knowledge quality seems to be a more important factor contributing to individual performance, team performance, and project performance. In social media online communities, knowledge quality is critical as the time and energy of project members to acquire knowledge are limited. Further, the project itself commonly has a tight schedule, which underlines the importance of knowledge quality.

Another limitation in current research on social media for project knowledge management is that much focus is put on social media technologies themselves. The adopted perspective regards social media as a discrete entity. As stated above, social media are adopted in a bottom-up characteristic. Different from traditional information technologies, managers know less about what really happens in some project online communities. Social media could also be used in unexpected ways. In this way, the socio-materiality perspective may be more powerful for phenomenon explanation in the project context (Ninan et al., 2020), thus appealing for more research.

References

Ahmed, Y.A., Ahmad, M.N., Ahmad, N., and Zakaria, N.H. (2019). Social media for knowledge-sharing: A systematic literature review, *Telematics and Informatics*, **37**: 72–112.

Azhar, S., and Abeln, J.M. (2014). Investigating social media applications for the construction industry, *Procedia Engineering*, **85**: 42–51.

Brooks, S., and Califf, C. (2017). Social media-induced technostress: Its impact on the job performance of it professionals and the moderating role of job characteristics, *Computer Networks*, **114**: 143–153.

Chang, C.Y., Pan, W., and Howard, R. (2017). Impact of building information modelling implementation on the acceptance of integrated delivery systems: Structural equation modelling analysis, *Journal of Construction Engineering and Management*, **143**(8): 04017044.

Chiregi, M., and Navimipour, N.J. (2016). A new method for trust and reputation evaluation in the cloud environments using the recommendations of opinion leaders' entities and removing the effect of troll entities, *Computers in Human Behaviors*, **60**(July): 280–292.

Cramton, C.D., Orvis, K.L., and Wilson, J.M. (2007). Situation invisibility and attribution in distributed collaborations, *Journal of Management*, **33**(4): 525–546.

Grant, R.M. (1996). Toward a knowledge-based theory of the firm, *Strategic Management Journal*, **17**(winter): 109–122.

Hasan, A., Ahn, S., Rameezdeen, R., and Baroudi, B. (2019). Empirical study on implications of mobile ICT use for construction project management, *Journal of Management in Engineering*, **35**(6): 1–12.

Huang, Y., Shi, Q., Pena-Mora, F., Lu, Y., and Shen, C. (2020). Exploring the impact of information and communication technology on team social capital and construction project performance, *Journal of Management in Engineering*, **36**(5): 04020056.

Jia, J., Ma, G., Jiang, S., Wu, M., and Wu, Z. (2020). Influence of social media use at work on construction managers' work performance: The knowledge seeker's perspective, *Engineering, Construction and Architectural Management*, Early Access.

Jia, J., Ma, G., Wu, Z., Wu, M., and Jiang, S. (2021). Unveiling the impact of task conflict on construction project performance: mediating role of knowledge integration, *Journal of Management in Engineering*, **37**(6): 04021060.

Joseph Garcia, A., and Mollaoglu, S. (2020). Individuals' capacities to apply transferred knowledge in AEC project teams, *Journal of Construction Engineering and Management*, **146**(4): 1–9.

Kaplan, A.M., and Haenlein, M. (2010). Users of the world, unite! The challenges and opportunities of social media, *Business Horizons*, **53**(1): 59–68.

Ko, D.-G., Kirsch, L.J., and King, W.R. (2005). Antecedents of knowledge transfer from consultants to clients in enterprise system implementations, *MIS Quarterly*, **29**(1): 59–85.

Leonardi, P.M. (2014). Social media, knowledge sharing, and innovation: Toward a theory of communication visibility, *Information Systems Research*, **25**(4): 796–816.

Leonardi, P.M. (2015). Ambient awareness and knowledge acquisition: Using social media to learn 'Who knows what' and 'Who knows whom', *MIS Quarterly*, **39**(4): 1–11.

Leonardi, P.M. (2017). The social media revolution: Sharing and learning in the age of leaky knowledge, *Information & Organisation*, **27**(1): 47–59.

Li-Ying, J., Zhang, Z., and Long, Q. (2018). An alternative way to make knowledge sharing work in online communities? The effects of hidden knowledge facilitators, *Management and Organisation Review*, **14**(4): 781–825.

Majchrzak, A., Faraj, S., Kane, G.C., and Azad, B. (2013). The contradictory influence of social media affordances on online communal knowledge sharing, *Journal of Computer-Mediated Communication*, **19**(1): 38–55.

Ma, G., Jia, J., Ding, J., Wu, M., and Wang, D. (2021). Examining the impact of social media use on project management performance: Evidence from construction projects in China, *Journal of Construction Engineering and Management*, **147**(3): 04021004.

McAllister, D.J. (1995). Affect- and cognition-based trust as foundations for interpersonal cooperation in organisations, *Academy of Management Journal*, **38**(1): 24–59.

Neeley, T.B., and Leonardi, P.M. (2018). Enacting knowledge strategy through social media: Passable trust and the paradox of non-work interactions, *Strategic Management Journal*, **39**(3): 922–946.

Ninan, J., Mahalingam, A., Clegg, S., and Sankaran, S. (2020). ICT for external stakeholder management: Sociomateriality from a power perspective, *Construction Management and Economics*, **38**(9): 840–855.

Okhuysen, G.A., and Eisenhardt, K.M. (2002). Integrating knowledge in groups: How formal interventions enable flexibility, *Organisation Science*, **13**(4): 370–386.

Oti, A.H., Tah, J.H.M., and Abanda, F.H. (2018). Integration of lessons learned knowledge in building information modeling, *Journal of Construction Engineering and Management*, 144(9): 04018081.

Pemsel, S., and Wiewiora, A. (2013). Project management office a knowledge broker in project-based organisations, *International Journal of Project Management*, **31**(1): 31–42.

Pivec, M., and Maček, A. (2019). Employment background influence on social media usage in the field of European project management and communication, *Journal of Business Research*, **94**(Jan.): 280–289.

PMI (Project Management Institute) (2017). *A Guide to the Project Management Body of Knowledge* (PMBOK guide), 6th ed., Newtown Square, PA: PMI.

Rambe, P., and Nel, L. (2015). Technological utopia, dystopia and ambivalence: Teaching with social media at a South African university, *British Journal of Educational Technology*, **46**(3): 629–648.

Reagans, R., and McEvily, B. (2003). Network structure and knowledge transfer: The effects of cohesion and range, *Administrative Science Quarterly*, **48**(2): 240–267.

Song, Q., Wang, Y., Chen, Y., Benitez, J., and Hu, J. (2019). Impact of the usage of social media in the workplace on team and employee performance, *Information & Management*, **56**(8): 103160.

Sundar, S.S., Oeldorf-Hirsch, A., and Xu, Q. (2008). The bandwagon effect of collaborative filtering technology, *CHI Conference Proceedings*, Florence, Italy.

Yu, L., Cao, X., Liu, Z., and Wang, J. (2018). Excessive social media use at work: Exploring the effects of social media overload on job performance, *Information Technology & People*, **31**(6): 1091–1112.

5

Social Communities of Practice
Reflection on Informal Mentoring in Curated Online Spaces

Elizabeth Harrin

5.1 Introduction

Professional networking once meant dinner meetings and after-work seminars, travelling to someone else's office or a shared meeting space in the evening to catch up with other people engaged in the same job. The connections made at events like these were good for sharing problems or discussing the challenges of project work. Individuals would also have the opportunity to network in the office, talking to peers in the project delivery community to share experiences and learn from each other. At home, people could offload some of the stresses of the day to the listening ears of family or housemates.

Those situations still happen, but increasingly, there is the need for alternative opportunities because the world has changed. More and more connections are made via mediated technology, more project team members are working remotely, and the COVID-19 pandemic has restricted in-person opportunities. As project management and related occupations gain widespread traction across a range of industries, more and more people are finding that they are working in project-related roles. In the UK alone, there are 2.13 million full-time equivalent works in the project management sector, supporting a profession that generates

Otobos Consultants Ltd, United Kingdom.

£1.56.5bn of annual Gross Value Added—a measure of total output and income of the PPM sector (APM, 2019).

However, many of these people may not be in large organizations with an internal network providing support and mentoring. In the UK, 99.3% of the private sector business population in 2020 was made up of small businesses, defined as having zero to 49 employees (BEIS, 2020). While some of these may have mature and robust approaches to project management training and professional development, it's unrealistic to assume that all small businesses would have the capacity to make that a priority for their team members. Indeed, a 2019 survey of global IPMA members and KPMG clients identified only 28% of organizations globally having a planned development pathway for project professionals and only 33% have a community of practice (IPMA, 2019). If organizations with the ability to join an international project management professional body or the budget to engage consultants for research can't make these opportunities available to every project professional, small and medium sized organizations are also likely to find it a challenge.

In addition, the chance of having someone at home to chat to is getting smaller. The workforce is increasingly made up of people who live alone. In 2020, the proportion of single-occupancy households in London was 22.8%. In Scotland, and the northeast of England, the proportion of one-person households was 33.6% (ONS, 2021). The number of people living alone in the UK is projected to increase to 10.7 million by 2039 (ONS, 2019).

Many project professionals lack access to an in-person community of practice or a safe space to talk about work issues, and so seek one out elsewhere—they are going online. While there are forums connected to professional bodies, these are typically only open to members and hosted on proprietary systems, which increases the mental effort required to seek them out and participate. Members have to log into a separate system, using the operating platform preferred by the institution, instead of using the social media tools where they are likely spend time anyway.

The solution may lie in curated communities of practice and peer-to-peer mentoring groups. Social media presents a broad opportunity for groups of project managers to come together in community spaces for informal support and knowledge sharing. Today's technology enables organizations and individuals to create spaces for informal social learning groups as a way to support continuous professional development of the project management community, whether that community is made up of employees only or members of the public. Everyone has a contribution to make, and a social learning community allows group moderators to tap into the wisdom of the crowd to support the wider community in a way that an individual PMO manager or a small corporate PMO team cannot.

However, social media platforms also present the opportunity for plenty of unstructured time-wasting. How can leaders create curated communities of practice that serve to support professional development in a cooperative and social way? To address this question, this chapter presents first-hand experience of facilitating two very different online communities: one on the social networking site, Facebook, and the other on the audio app, Clubhouse. Subsequently, the chapter discusses four foundations for creating a successful community, along with practical tips for engaging members of the community so that participation and energy in the group are high. Next, the chapter offers some tips for community moderators and facilitators to ensure a balance of time spent between community engagement and other work. The chapter concludes with a brief look at the future direction of online communities.

5.2 The Community Experiments

The author has been a member and administrator/moderator of many online communities for various different purposes. The focus of this chapter is the opportunity for informal, co-created mentoring in social spaces, and the community groups that have leant themselves best to this end have been the ones the author has created on Facebook and Clubhouse. Other social platforms, like LinkedIn, do allow for the creation of groups, but not all social media channels have the ability to easily bring large groups together for the purposes of social learning. This chapter looks at the author's experience of curating social spaces for peer-to-peer mentoring on two channels that do provide that opportunity.

5.2.1 Facebook: The Project Management Cafe

Facebook is a social networking site with 2.85 billion monthly active users (Facebook, 2021). With the exception of YouTube, Facebook is by far the social platform with the widest reach, with around 69% of adults reporting using it (Auxier and Anderson, 2021). The majority of Facebook users say they visit the platform on a daily basis (ibid.), so social communities that use Facebook as their 'home' are likely to reach a wide section of the project profession. Users have the ability to create public and private groups that serve a variety of purposes. For example, groups can be created to bring together local communities, to create a forum for customers of a business or users of a product, to bring together attendees for a specific event, or for topic-based discussion, where users with similar interests are able to connect and engage with each other in the space.

The Project Management Café group was set up in February 2017 to serve as the latter—a topic-based discussion group for project

Table 5.1: Facebook Group Membership for the Top Five Countries.

Country	Members
United States	1,930
United Kingdom	709
India	264
Canada	183
Australia	129

professionals. At the time, much of the community-driven groups and content on Facebook were around exam preparation and test taking. There was space for a community of practice that was aimed at working project managers—a space that could be a permanent home for professionals instead of something they joined while they studied for an exam. At the time of writing, the Project Management Café group has 5,160 members spanning over 95 countries. Membership from the top five countries is shown in Table 5.1.

5.2.2 Clubhouse: The Project Management Club

Clubhouse (https://www.joinclubhouse.com) is an audio-based social community app. Designed as a 'space for authentic conversation and expression' (Clubhouse, 2021), it's a drop-in audio platform that allows users to join in any conversation and take part in the discussion at the host's discretion. It can be considered as a large conference with multiple rooms. Participants enter any room to hear the speaker. Sometimes the speaker is a single person giving a presentation; sometimes they accept questions from the audience. Other rooms may play host to a panel debate. Most rooms in Clubhouse allow for audience members to raise their hand and come up to the 'stage' to ask a question or contribute their own experiences to the discussion. This makes it a very interactive and engaging experience but as rooms cannot be recorded, the participants have to be there live to hear the conversation. In the autumn of 2021, Clubhouse added the ability for users to record snippets, called 'clips', of up to 30 seconds. So, it remains to be seen whether this functionality to record and broadcast rooms after the event will be extended in due course.

The app allows users to create their own 'clubs'. These groups make it easier to find topics that users are interested in and to curate a feed of relevant conversations to join. Group members have additional privileges, such as being able to start conversations ('rooms') within the app at the group owner's discretion, and they can opt to receive notifications about scheduled upcoming discussions. The Project Management Club was set up in February 2021 and had 20,800 followers at the time of writing in

July 2021. Clubhouse feels like it has caught a wave of engagement and has tapped into something people feel they need: human conversations in real-time on a casual basis.

5.2.3 Informal Mentoring in These Communities

Both the Facebook group and the Clubhouse club have the same aim: to connect professional project managers and those wanting to grow in the profession to provide informal support—the same kind of support that one would get from coffee with a colleague at work or by chatting to someone in the line for food, at a conference. That informal mentoring and connection helps to build confidence and get questions answered—especially questions an employee would not want to ask his/her boss.

In the Facebook group, community members ask questions of each other and often many people contribute their experiences or thoughts to the debate, sharing examples or resources to help the original poster. This interaction is not in real time and the algorithm dictates what appears on the user's feed at the time they open Facebook. Questions may get missed, depending on what else Facebook considers is worth presenting to the user; typically, the more engagement a post has, the more likely other people are to see it. The moderators share questions anonymously on behalf of members who do not want to be associated with the question, and also use conversation prompts and other structured engagement activities to keep interactivity high. From time to time, there are live video training sessions, though typically these are not scheduled in advance and interaction in real time is low.

In the Clubhouse group, community members join a scheduled room at a particular time, either because they plan to be there live or because they happen to be using the app at that time and are attracted to join the conversation by the room title. Rooms can attract around 150 people, a dozen or so of whom will want to ask a question or join the discussion; the others are present in the audience to listen to the debate. Participants come from all over the world and a variety of backgrounds and from highly experienced to just starting out in their careers. The time the room is open plays a large part in determining the geographical location of the audience members; it is important to host rooms at a time when the participants are available to take part, which is typically before or after working hours. Types of commonly asked and answered questions in both groups are shown in Table 5.2.

There is a distinct difference between the types of questions asked. Asynchronous conversation and co-created answers in Facebook create a far richer discussion about the soft skills required to successfully operate as a project manager. The Facebook engagements often talk about frustrations and challenges, and the nuances of leading a team in a matrix

Table 5.2: Themes from Commonly Asked Questions in the Social Communities.

Facebook: Project Management Café	Clubhouse: Project Management Club
Project management software recommendations	Understanding job requirements
Tips for stakeholder engagement	Help choosing a certification
Managing matrix teams	Tips for moving industries
Taking over a project	Tips for moving into project management from another role, e.g. business analysis, development
Recovering a troubled project	

environment, often with a lot of different internal political influences at play. In Facebook, the topics feel more substantive to project management practice and the day-to-day responsibilities of doing the job. However, the responses are typically only a few sentences and lack the depth that would be expected in an in-person conversation with a mentor. In the author's experience, conversations on Clubhouse are deeper and the connections have the potential for a greater impact due to the fact that the interactions are happening in real time. Word for word, interlocutors are likely to get a longer response to their question on Clubhouse with more input, more tailored to their personal situation. People asking questions also tend to provide more context and there is the possibility for real-time clarifying questions. It's usual for participants to ask two or three related questions in one go whereas Facebook users tend to ask clear, direct questions relating to a specific challenge.

The Clubhouse conversations facilitated by the author in a weekly, 'Ask a Project Manager' room, are far more predictable, and it's common to get career-based questions. As the format of these discussions is more of an 'ask the expert' style, there is less community interaction for each question. However, others in the room do occasionally offer their experience to another person, or the author or room host can call on someone else to share their views when it is recognized that they would have something of value to add.

Both these communities offer the opportunity for informal mentoring and connection with other professionals: a kind of water cooler environment for chatting with others in a similar role and canvassing opinions on the topics that are important. This interactive chat plays a crucial part in a projects manager's professional development, both in the ability to ask questions and to offer an opinion that the other person values. However, neither of the social groups discussed above completely replaces the value of an in-depth conversation with a trusted peer or mentor who fully understands the circumstances and background. When the author holds that type of conversation with project managers, typically the goal is to build an ongoing relationship and have the time to explore options, create plans, and take action to address the challenge in a way that can be monitored, tracked, and assessed over time to ensure that goals are met. Social communities provide a more 'bite-sized' and peer-led approach to informal mentoring.

5.3 Creating a Successful Community

This section considers the lessons from facilitating social learning communities on both the Facebook and Clubhouse platforms with a view to identifying how these can be applied to other communities of practice within a PMO setting, project management professional associations or user-created communities that coalesce around a topic.

The first step in creating a successful community is to define what success means for that community. What is the purpose behind the community, and what does it exist for? How should community members show up and participate? For example, one goal for a social learning community may be around sharing lessons learned within a multinational organization, or to provide a platform for co-created video training. Perhaps the objective of the group is simply to create a space where project managers can lean on each other when times get hard, and share tips and resources. If the goal is clear, members of the community will find it easier to find their place and know how they are expected to participate.

5.3.1 Making the Time

A key success criterion for a thriving community is investing the time in nurturing the group. The author's experience has shown that this does not need to be extensive. But one cannot expect a community to grow and be useful to each other without a little overarching support from the host or moderator. Organizers and community hosts should plan to spend time curating content for the group so that if there are no member contributions, participants can still engage with each other on pre-prepared content. It is important to create a cadence for the group so that members know what to expect and where to spend their attention. For example, in the Project Management Café, Tuesdays are sharing days where members can share their own resources or links to materials they have created. The rhythm of the community can unfold or change with time as the group gets to know each other. As the community develops a culture of its own, the group moderators can adapt and flex their own involvement to meet the needs of the group.

5.3.2 Facilitate Proactively

Proactive facilitation means making sure only relevant conversations are taking place in your community. It is important to make it clear to participants that they are being watched, or heard, so that individuals feel that there is a level of moderation within the community. On Facebook, that means deleting inappropriate threads or turning on comment moderation for people whom the moderators believe may interact inappropriately. In some cases, that means blocking people from the group. In Clubhouse, it means consciously curating the stage by checking a user's profile before inviting him/her to speak and being quick to remove people from the stage or report them for disruption, if necessary. In both cases, it means making connections between users and joining the dots for people, pointing out resources, and tagging or mentioning individuals who are able to provide further support.

Online facilitation and 'live' facilitation are very different skills. Moderating an online, text-based social community can be done at any time. Facilitating a live, drop-in audio room requires the ability to bring in contributors in a timely fashion and keep the conversation moving and is a good opportunity to develop public-speaking skills. Both settings require the community host to shape the direction for the group and set the tone for interactions. Effectively setting the tone first means that group moderators must consider what a good result for facilitation would look like in that community and what it would take to achieve that.

5.3.3 Set the Ground Rules

Part of setting the tone for interaction is creating ground rules or social norms for the community. The rules for Project Management Café are a mixture of culture-setting and guidance on how to interact in the group. The rules in operation at the time of writing are informal and accessible, including directives against hate speech and bullying, clear guidance against posting materials that could be considered spam or promotional, a request to be kind and courteous, and a call to respect the privacy of other members. People are asked to agree on the rules on requesting to join the community, and in the main, they do complete the entry questions. Many members feel protective of the space and are quick to flag and report inappropriate content if they see it. However, a frequent occurrence is members agreeing to the rules and then promptly breaking them within hours of joining the community. In that case, they are either removed or their ability to post in the group is limited.

Similar rules are applicable in the Clubhouse Project Management Club community, namely staying on topic within rooms, and being kind, courteous, and respectful. There is an additional point listed in the club rules around staying connected to the community through other social channels, but this is not enforced and is more of a note to prospective members about where else they can find discussion when there are no club rooms active.

Group organizers and moderators should consider how ground rules will be created and enforced if the community is open to the public. This will enable the moderation team to have a plan in place when individuals choose to break the rules. Internal, corporate communities made up of staff members, will, in the author's experience, require less effort to enforce ground rules as there are fewer incidents of rule-breaking, and incidents may potentially be less impactful. Moderators may still consider it a worthwhile exercise to co-create a list of expectations and norms for how community members should behave while participating in the community.

5.3.4 Reduce Barriers to Participation

Creating an engaged community of practice relies on having a membership that is engaged. That starts with operating on a platform that makes it easy for people to get involved. Consumers—your potential community members—are spending more time online but that doesn't mean they enjoy it (Lai, 2021). Making it easy to get involved will reduce the technology fatigue that comes with any kind of online activity. As a rule of thumb, go where the people are; don't expect the people to find your community if it is created in a location where the target audience is not active. For social media-led groups, which are open to the public at large, that means creating a group on someone else's platform. There are risks with 'digital sharecropping': building a community on land belonging to someone else (Carr, 2017). Rules can change and groups can be deleted without warning. In addition, some community members will not want to engage with platforms where concerns have been raised about data privacy, which therefore limits the community membership to only users who feel comfortable using those platforms. However, those risks need to be offset against the low barriers to entry for many participants. Community moderators will need to decide if 'many' users would be a better outcome for the social learning group than 'all' users and this will most likely depend on the target group for membership.

PMO practitioners may take the decision to create a community of practice for the organization using in-house tools that are already present in the business, for example, using a forum or Slack channel. If the target audience of employees is already actively using the platform, then it will reduce the level of friction for them to engage in the community. Whatever tool is chosen, look for what can be done to reduce the barriers to participation, for example, removing the need to log in to a separate platform by using single sign options or a tool where they are automatically logged in with cookies in their browser. Moderators should consider what can be done to make it easy to participate in the discussion by keeping questions brief and specific. For example, instead of asking people to share a detailed description of a challenge and corresponding lesson learned on the project, ask for their top tip in a particular situation. Members will proactively share other things on their own once they are familiar with the community and how it works. If the expectation is that users will co-create responses in the space for peer-to-peer mentoring, then make sure group members have technical permissions set to be able to do that and are notified when relevant posts are created that align to their areas of expertise and interest. Using a friction-free tool that is already known to the user base will encourage participation and make it easy for people to show up in the community.

5.4 Practical Tips for Community Engagement

Setting up a community is only the start. Once the group is set up, goals or objectives for your community are agreed, cultural expectations for members are co-created, and moderators are aware of the time and engagement requirements, then the work really begins to bring people into the community and keep them engaged while they are there.

In a public social media-led community, not everyone will be active. The Project Management Café group has around 1,200 active members on the most active days and this represents about a fifth of the total membership. Inactive Facebook users are proactively removed so the group does not carry members who are no longer using their Facebook account. However, many members will simply prefer to read instead of engaging with the content: Li and Bernoff (2011) call these groups Joiners or Spectators. It will be the Conversationalists and Creators that will actively participate. So how are people encouraged to engage with the online community? The sections below share some experience-driven tips for doing exactly that.

5.4.1 Be Visible

Humans prefer to connect with humans, so encourage community members to upload a profile photo that shows their face. They should use their real names and there should be no opportunity for anonymous posting, unless done so via a moderator to preserve the identity of the original poster. Where appropriate to do so, share 'behind the scenes' photos so that community members are encouraged to recognize each other as colleagues. For example, within the Facebook group, the moderators ran a scavenger hunt where people shared photographs of different office-related items every day, including a picture of their workspace. If there is the option for engagement with real-time video, encourage participants to have their cameras on. Audio-only social groups, like Clubhouse, rely on people to show their personality through their profile picture and bio, and also through their links to other social platforms where members can showcase photos.

5.4.2 Make Content Visible

As well as being personally visible as a human within the community, it is important that both moderator-created content and user-created content are easily findable. Use the features within the social platform to organize content. Hashtags, topics, tags, guides, units, and other features make past content searchable, which is important for professional development as answers will be available at the time they are needed. It also limits the repetitive nature of similar questions coming up time and time again. It's highly likely that members search for a question or to see if someone else has had a similar challenge to themselves before reaching out to the community for support, and moderators can encourage that too. It saves time for members and the moderation team if it is easy to find the past discussions; for example, create transcripts of video sessions so they can be searched by keyword.

5.4.3 Ask Questions

The simplest type of content to create for any community is to ask a question. This can be in the form of a text-based post, an image, a poll, a video or during a live interaction with community members. Questions prompt answers, and answers mean people are engaging with the content. When curating content for the community, focus on topics that allow members to talk about their successes and skills. People are more willing to share positive experiences than negative ones. In other words, getting people to share a win or something that allows them to show their skills in a positive light will encourage more engagement than questions that ask for challenges or anything that could be construed as a confession.

Vary the type of response needed so that even people short on time have the opportunity to respond to some content; for example, make it easy for people to answer questions by providing a poll-based response with a set of preselected answers to choose from. A participant only needs to choose an option to engage with the community content in that case. Other question-based content can inspire longer form responses for those that choose to engage with it. Where members ask questions, make sure there is a response; never leave a question unanswered. This may involve the moderator responding or bumping the question to the top of the feed so that other people are more likely to see it. Group admins can also reach out to qualified members and ask them to comment on the post, thereby providing a response. Community members want to know that someone is listening and that they will get the support they need in a timely manner. The more people see that this is happening, the more likely they will be to ask their own questions and seek out the informal mentoring and peer-to-peer debate that enriches online communities. Moderators should make it a habit to regularly check for member-generated content without responses so that some kind of response can be provided.

5.4.4 Make New Members Feel Welcome

New members of the community may prefer to watch and learn from others to get a feel for how the group operates and what is expected of them. Community organizers can welcome new members and encourage them to take a small action. Many groups invite new members to introduce themselves, although this technique is not always successful. Some members may feel comfortable doing that straight away; others may not want to put themselves in the spotlight on first joining. If the community is made up of employees, then expecting them to introduce themselves presents less of a barrier. A manager or colleague can introduce the new starter and welcome them to the group.

Within a social media context, moderators should be prepared for members to not introduce themselves even if they have been invited to do so. An alternative approach is to create a welcome series that provides new group members with an orientation to the community. The Project Management Café group uses a three-part orientation series. Members are tagged in a welcome post and invited to click through a series of posts to learn more about the community. Each post includes the links to the other orientation posts. Clicking on several posts within the group also helps the platform recognize that this group is important to the user. That may influence what content Facebook chooses to present to new members which in turn, supports higher levels of engagement on the platform.

5.4.5 Be Consistent

Communities thrive on consistency. Whether it is a weekly poll, a monthly curated selection of top articles in the space that the community would appreciate or showcasing a resource from the PMO once a fortnight, it is important to create a routine and stick to it. The greatest challenge the author has seen for community groups is that they are created with a large amount of buzz and fanfare, with high engagement from everyone involved. However, that is not sustained due to the founder and/

or facilitators having other commitments or failing to prioritize their community engagement work.

Many social channels allow group owners to schedule non-live content, so it is possible to plan content ahead of time when moderators are not physically going to be present to do it. Engagement-led discussion topics and resources can be scheduled in advance to maintain the cadence of the group and spread out the content. This will provide a better experience to group users than sharing five or six pieces of content in a day and then nothing for the next two weeks. It's also worth considering what backup plan should be in place for times when organizers are unable to actively moderate the community. One option would be to plan (and announce) that the group will be less active over the summer holidays or end-of-year festivities. Alternatively, some groups work well with a pool of community champions who can step in when other commitments mean the primary facilitator is unable to carry out their normal activities in the group.

5.4.6 Do More of What Works

Most platforms provide statistics and data on the membership. It's important to review community engagement data and look at what is working for the community so that appropriate action can be taken. Within the Project Management Café group, live video gains more engagement over time than other content, and text-based question posts that ask contributors for their advice in a given situation also encourage members to contribute. These get many long-form comments where members weigh in to share their expertise and experience with the person asking the question. In Clubhouse, drop-in question-and-answer sessions generate greater turnout and more contributors than interviews or topic-led discussions. These results are highly specific to the nature and makeup of the groups, and what's popular in other communities will be different. Note that holidays and other seasonal influences will also affect the engagement levels within the community. When reviewing data about participation or engagement, moderators should take into account seasonal changes based on national and international events.

Group organizers can review any available statistics and use their professional judgment to uncover what is working for the community. Moderators can also survey members about the kinds of content they would like to see. Ultimately, the community exists to serve the members, and the members co-create the experience of the community by virtue of being in it. This might mean that the community goals evolve over time as the moderation team or community champions double-down on what is effective and useful for the membership.

5.5 Managing Your Time

Participation in any community takes effort and energy, and that needs to be balanced with other responsibilities, such as getting projects delivered; it's not practical or desirable for anyone in the community to spend too much time on social channels, however useful they are for professional development and peer-to-peer support. As a moderator of online spaces, schedule time to carry out community engagement work and make use of technology to pre-plan where possible. A sustainable cadence for the group should be created; for example, live video training every day might sound good but after a couple of months, those in front of the camera will be exhausted and probably struggle to come up with new topics. It's always easier to add events for the community into the schedule, instead of taking them out. If the community includes live events, make sure these are programmed ahead of time and that people have a way to find out when they are taking place. Organizers can promote live events to community members and, if appropriate, to people outside the community as a way of driving participation in the event and increasing the membership. Checklists or a documented process for creating, hosting, and post-event activities can save time. Once the moderator has a format that works, turning it into a repeatable process will also allow the work to be delegated to community champions or other administrators.

A community that is 'always on' is helpful for members in different time zones, e.g., international business teams. However, the community administrator does not need to be present at all times. If necessary, reserve 15 minutes each day for moderation activities: approving new member requests, reviewing member-flagged content, and quickly checking for spam comments. The more the group is encouraged to self-police and report inappropriate content, the easier this will be for the administrators. Community moderators can let the membership know when they will be available or are usually online in case members want to get in touch in real time.

Templated responses to common member questions are another way to save time, along with using FAQ (Frequently Asked Questions) or the group description and rules to share the basic expectations about community engagement. Aim to reduce the number of common questions received about how the group works or what should be shared when, by making this information easily accessible. If moderating an employee-based community is a responsibility on top of the administrator's day job, it may be worth keeping records of the time spent on this activity. As the community grows, it may be necessary to employ a full-time moderator or reach out to the community to appoint some champions to help share the workload.

5.6 Future Directions

The data and social channels mentioned in this chapter will no doubt date with time; possibly very quickly. The world of online interactions changes frequently as new tools are launched and others consolidate or close. At the time of writing, several other social channels were trying out audio-only spaces to boost interaction and create the environment for engagement in a way similar to Clubhouse. As a result, it's impossible to predict with certainty where social media and curated online communities might go in the future. However, it is realistic to assume that hybrid workplaces and remote work are here to stay. Virtual working, offshore and nearshore teams, collaborative contracting, and other ways of curating teams from many organizations to collaboratively deliver projects: all these ways of working mean it is more and more likely that colleagues will not share office space with the people who can help them best. That makes it even more important to find ways to connect socially and professionally with colleagues, and it follows that technology will continue to mediate those connections.

Research by GWI in 2021 shows that the pandemic shifted how people chose to use social media. Up until the end of 2019, social media usage was more about consuming content. A shift was identified in 2020 that saw users seeking out community connection through social media, reminiscent of how many of the social channels first grew a following: by encouraging users to stay in touch with friends and family, and to share their personal news. 'Keeping in touch' is the main reason that 50% of users log into social media, with work-related networking being the top reason for only 23% of users (GWI, 2021).

The kind of communities of practice represented by social learning spaces online do not replace the power of in-person connectivity and sustained workplace relationships that build over time, and those should also be fostered and encouraged, where possible. However, it seems like there is a trend towards the desire for deeper connectivity and the ability to have 'real' conversations in a setting that acknowledges and appreciates the complexities of work/life balance. Authenticity, and the willingness of others to be their authentic selves in a way that is not judged, is an expectation of the moment. It is possible to do that online, if the right spaces are curated.

There is huge value in curating a community that supports each other and provides informal peer-to-peer discussion and learning in a social context. As experienced people continue to reach the age of retirement, new employees need access to knowledge management systems that help prevent 'brain drain' from organizations. Tapping into the wisdom of the crowd from within and outside your organization by creating spaces for informal peer-to-peer mentoring is one way to support continuous professional development for the project management community.

References

APM. (2019). *The Golden Thread: A Study of the Contribution of Project Management and Projects to the UK's Economy and Society*, London; Association for Project Management.

Auxier, B., and Anderson, M. (2021). *Social Media Use in 2021*, Pew Research Centre. [accessed on 7 July 2021]. https://www.pewresearch.org/internet/wp-content/uploads/sites/9/2021/04/PI_2021.04.07_Social-Media-Use_FINAL.pdf.

BEIS. (Department for Business, Energy and Industrial Strategy) (2020). *Business Population Estimates for the UK and Regions 2020: Statistical Release* [accessed on 6 July 2021]. https://www.gov.uk/government/statistics/business-population-estimates-2020/business-population-estimates-for-the-uk-and-regions-2020-statistical-release-html.

Carr, N. (2017). *Utopia is Creepy and Other Provocations*, New York: Norton.

Clubhouse. (2021). *Community Guidelines* [accessed on 6 July 2021]. https://www.notion.so/Community-Guidelines-461a6860abda41649e17c34dc1dd4b5f.

Facebook. (2021). *Facebook Reports First Quarter 2021 Results* [accessed on 6 July 2021]. https://investor.fb.com/investor-news/press-release-details/2021/Facebook-Reports-First-Quarter-2021-Results/default.aspx.

GWI. (2021). *Social Media Marketing Trends in 2021* [accessed on 7 July 2021]. https://www.gwi.com/reports/social.

IPMA. (2019). *The Future of Project Management: Global Outlook 2019* [accessed on 7 July 2021]. https://www.ipma.world/assets/PM-Survey-FullReport-2019-FINAL.pdf.

Lai, A. (2021). *A Consumer 'Tech-celeration' is Advancing Behaviors but Straining Emotions* [accessed on 7 July 2021]. https://go.forrester.com/blogs/a-consumer-tech-celeration-is-advancing-behaviors-but-straining-emotions/.

Li, C., and Bernoff, J. (2011). *Groundswell: Winning in a World Transformed by Social Technologies*, Boston: Harvard Business Review Press.

ONS (Office of National Statistics) (2021). *Families and Households in the UK: 2020* [accessed on 6 July 2021]. https://www.ons.gov.uk/peoplepopulationandcommunity/birthsdeathsandmarriages/families/bulletins/familiesandhouseholds/latest.

ONS (Office of National Statistics) (2019). *The Cost of Living Alone* [accessed on 6 July 2021]. https://www.ons.gov.uk/peoplepopulationandcommunity/birthsdeathsandmarriages/families/articles/thecostoflivingalone/latest.

6

Investigating the Role of Social Media on Megaprojects

Sunila Lobo

6.1 Introduction

There has been an increase in megaprojects around the world, including the Beijing-Guangzhou high-speed railway, the US$23 billion expansion of the world's first underground railway, which is the London Crossrail project (Desjardins, January 2017), and HS2, which is a new high-speed railway with the aim of linking up London, the Midlands, the North and Scotland (*Source*: hs2.org.uk). Costly, above US$1 billion and complex, megaprojects attract massive public attention due to a huge impact on the environment, economics, communities, and societies (Altshuler and Luberoff, 2004; Batheram et al., 2005; Giezen, 2013) and at times, are the cause of conflict from the public and other stakeholders (Flyvbjerg, 2014).

Stakeholders are defined as '*any group or individual who is affected by or can affect the achievement of an organization's objectives*' (Freeman, 1984; Freeman et al., 2010). In fact, megaprojects can involve multiple stakeholders with conflicting interests (Biesenthal et al., 2018). Stakeholders are divided into external and internal stakeholders (Bonke and Winch, 2002) where members of the project coalition are internal stakeholders. Those who are affected by the project in a significant way are external stakeholders, while the affected members of the public are public stakeholders.

Research has shown that stakeholders have the ability to influence a project and its success (Oliveira and Rabechini, 2019; Lobo and Abid,

Worshipful Company of Marketors, London, United Kingdom.

2020); thus, effective managing of the communication and relationship with stakeholders is important. The use of technologies like social media can provide 'voice' for the affected local community and potentially influence project promoters' decision-making behaviors (Lobo and Abid, 2020). Project stakeholder management has been transformed in the social media era, providing both opportunities (Ninan et al., 2019b and 2020b; Mathur et al., 2021) and challenges (Williams et al., 2015), with some public stakeholders using peaceable and formal means (Lobo and Abid, 2020) in the hope of strengthening their legitimacy as opposed to resorting to aggression or riots (Dewey and Davis, 2013; van den Ende and Marrewijk, 2018) to make public their dissatisfaction. The next section briefly re-views the relevant literature.

6.2 Literature Review

The importance of external stakeholder management to project success, particularly for infrastructure megaprojects, has been increasingly recognized (Aaltonen et al., 2008; Aaltonen and Kujala, 2016; Bourne and Walker, 2005; Di Maddaloni et al., 2018; Fewings, 2013; Jepsen and Eskerod, 2009; Olander and Landin, 2005; Olander and Landin, 2008).

There are power perspectives in external stakeholder management (Ninan et al., 2019a; Ninan et al., 2020a), with social media positioned as a critical ICT tool for external stakeholder management in infrastructure megaprojects (Ninan et al., 2020b), including for organizational legitimacy (Suchman, 1995). Ninan et al. (2019b) developed six propositions, through conducting content analysis of tweets by the infrastructure organization of a metro rail project in India, and the community's responses; augmented with interviews with project teams. Progress updates of the project comprised 61% of all tweets; to keep the community informed and to be assured of safety measures taken, and there were also tweets on promotional events and programmes to raise awareness. Targeting specific sections of the population, the metro rail project organization also made clear its social commitment and sensitivity to national and regional sentiments, which altogether had a positive effect on the project team and the community, some of whom became advocates for the project (Ninan et al., 2019b). However, this is in a particular context in India and the propositions would need to be tested elsewhere.

There is also a growing stream of literature on the benefits of using social media for project management (Daemi et al., 2020), including exploring pre-adoption and post-adoption effects (Ma et al., 2021), for transport planning and operation (Nikolaidou et al., 2018), and for benefits realization in transport projects (Mathur et al., 2021). In addition, it is even used for disaster management (Shandraseharan et al., 2021) and for better decision making; through social media data analytics (Tang et al., 2017). For the latter research, set in the US construction industry, Twitter posts of four clusters (construction workers, construction companies, construction unions, and construction media) were analyzed, finding that most negative posts were from the construction workers' cluster (Tang et al., 2017). This shows that decision makers need a better understanding of the actual situation in the industry, with Twitter being one of the information sources (Tang et al., 2017). Investigation of such social media use in the construction sector could have some useful findings; for instance, on benefits such as improving information sharing, engagement and relationships in the sector (Daemi et al., 2020), but this is not the focus of this chapter.

The focus is the online and to a lesser extent, the offline strategies of stakeholders to influence decision making on infrastructure megaprojects, including coercion and compromise (Frooman and Murrell, 2005). Project organizations' five strategies to respond to stakeholders' claims (Aaltonen and Sivonen, 2009) are adaption, compromise, avoidance, dismissal, and influence. While stakeholders' eight influencing strategies (Aaltonen et al., 2008) during the project execution stage are resource building, credibility building, direction action, coalition building, communication, conflict escalation, direct withholding, and indirect withholding.

According to Yang et al. (2011), the process of identifying stakeholders and their needs, categorizing and prioritizing stakeholders because of the assumption that some among them matter more than others (Derry, 2012), assessing their impact, and formulating effective engagement strategies is the main challenge that project managers face in megaprojects. Stakeholder trust formation towards the focal firm can be impacted by this process (Hayibor and Collins, 2016; Pirson et al., 2017; Pirson et al., 2016), with Cheng et al. (2017)'s development of an Onion Cognitive Model for trust influencing factors in social media communication—an interesting model to investigate for the project context. This model is based on a qualitative study of WeChat communication (Cheng et al., 2017).

It is critical for project managers to understand the mobilization strategies of stakeholders, so they are able to predict their behavior and thus, manage stakeholders more systematically. This challenge becomes more complex in the age of social media because of the proliferation of communication, especially as project benefits are perceived differently by citizens and project promoters (Mathur et al., 2021). However, social media, if harnessed, is perceived as a critical ICT tool for external stakeholder management in infrastructure megaprojects (Ninan et al., 2020b).

In their paper on Crossrail 2, Lobo and Abid (2020) therefore attempt to fill the gap in the literature by providing a 'stakeholder actions' perspective. The aim of these stakeholders who oppose projects is to bring about changes in the activities of the focal firm/project promoter, which can result in costs, and thus is a critical issue for managers (Hayibor and Collins, 2016; Teo and Loosemore, 2014). Investigating intrastakeholder structures and the possible influence of these structures on stakeholder action (as recommended by Williams et al., 2015) was carried out in the context of public stakeholders' actions including the use of social media to mobilize affected people about the impact of a proposed new Crossrail 2 station on the King's Road in Chelsea, London. This focus provides a contextualized perspective, as advocated by Mok et al. (2015).

Social media allows easy connections, sharing of messages, and organizing protests, with minimal resources. This reduces the costs of forming a stakeholder alliance, enabling weaker stakeholders to organize action. This is an incentive for their mobilization (Frooman, 1999). Lehtinen (PhD thesis, 2021), examining social media as an engagement practice in the Rantatunneli project in Finland, found that project organizations engage social media stakeholders successfully by being cooperative and communicating topical information about the project's progress (as in Ninan et al., 2019b) and sustainability issues. The latter issue, which is about environmental sustainability, is the reason for the opposition towards HS2 as it passes through ancient biodiverse woodland, i.e., Jones Hill Wood, near Wendover in Buckinghamshire.

This chapter analyzes Twitter data related to Jones Hill Wood and HS2, in order to understand the interactions between various stakeholders and HS2, as they pursue seemingly opposite paths—sustainability of the environment versus sustainable travel. The following section briefly describes the methodology used.

6.3 Research Setting and Methodology

Naturalistic inquiry is a qualitative research method used to examine people's life experiences in context (Lincoln and Guba, 1985). In project management research, however, qualitative research is still largely limited to gathering data through interviews, rather than investigating naturally occurring data, for instance online (Ninan, 2020). However, in Kozinets' netnography (2010; 2014), the context investigated is online; for instance, through Instagram and Facebook. An example is the investigation of online social networking by diabetes patients on Facebook (Greene et al., 2011).

An exploration of online naturalistic inquiry is important for understanding projects and project settings in the 21st century; for instance, interactions with both internal and external stakeholders (Ninan, 2020). In the approach taken here, the researcher is passive, with no direct interaction nor involvement with the online community, as described in Dekhoda et al. (2020)'s observational netnographic investigation of comments about dementia in five open Facebook communities. Despite its advantages of being a non-interventionist approach, online naturalistic data does not represent tone of voice or non-verbal cues, thus making it difficult to discern whether people are being serious, sarcastic or making a joke (Germain et al., 2018).

This chapter draws on the case study of the consultation phase of the Crossrail 2 megaproject in Chelsea, London, focusing on two public stakeholder groups (and intragroups) and their engagement on social media (Lobo and Abid, 2020). This would help further to advance research by examining the strategies used to oppose a particular 'intervention' by HS2 as it passes through ancient woodlands in Jones Hill Wood, in Buckinghamshire.

The first step in the research process for the Crossrail 2 megaproject case (Lobo and Abid, 2020) was to review the literature, identify the gaps, and devise the research question. This was followed by developing the first interview protocol to determine industry experts' views on social media strategy for stakeholder management. Social media content analysis of public stakeholder groups was also conducted. Themes were derived inductively on strategies used by public stakeholders to mobilize and influence decision making. These themes were used to develop Interview Protocol 2 for use with the opposing group of public stakeholders. The purpose of these interviews with public stakeholders was to develop an in-depth understanding of intra-groups of public stakeholders, their interests, and strategies. This helped determine their collective identity development and action. It also allowed for triangulation of data obtained from the social media content analysis.

Table 6.1 compares some of the basic details for both cases, and the differing focus of the external stakeholders opposed to these two UK infrastructure megaprojects. The focus of this chapter is external stakeholder opposition on social media, to the destruction of the natural environment in Jones Hill Wood including ancient trees and rare species, like barbastelle bats, in HS2's path. Secondary data from websites and news sites are also used to triangulate against the online social media data. Twitter posts opposing the removal of habitats and ancient trees from Jones Hill Wood (JHW) by some of the external stakeholders are analyzed and discussed in the next sections.

Content analysis was used to analyze the use of social media by external stakeholders to mobilize and influence decision making on the proposed new Crossrail 2 station in King's Road, Chelsea.

There has been vociferous local opposition on social media to the proposed station site in Sydney Street, near the Old Chelsea Town Hall. Public stakeholders from the NoCrossrailChelsea group, including

Table 6.1: Comparison of 2 UK Megaproject Cases where External Stakeholder Actions are Examined.

	HS2	**Crossrail 2 (Lobo & Abid, 2020)**
Location of case	Ancient woodlands at Jones Hill Wood.	Proposed new station at King's Road, Chelsea
Reason for opposition to project	Felling of trees and the destruction of natural habitat of rare species.	Cost/waste of taxpayers' money, Disruption to village character, local businesses and potential damage to historical buildings and homes on King's Road.
External stakeholders opposed to the project	Local and national opposition, environmental groups, activists, wildlife and conservation organizations.	Local opposition: residents, local businesses, Chelsea Society, etc.
Date focal firm joined Twitter	@HS2ltd joined in March 2011.	@Crossrail2 joined in March 2013.

celebrities living in Chelsea, like Felicity Kendal, used Transport for London (TfL)'s 2015 announcement of its plans to increase tube capacity, as proof that a new station is not needed in King's Road, Chelsea. Royal Brompton NHS Hospital chiefs also voiced their objection to the new station because it will 'seriously undermine' its plans for a new state-of-the-art centre to treat chronic lung disease patients. This is because the station site and access shaft footprint are on land planned for sale by the hospital to fund new treatment facilities (*Evening Standard*, 27 October 2015). As mentioned in Table 6.1, the proposed station construction may cause damage to houses and listed buildings along the proposed route. This is because it would require boring of tunnels which are only 20 metres down and would pass under 300 listed buildings along the proposed route. The fear of vibrations and their impact on the structure of residents' homes, is a major concern that was expressed in interviews with the opposing group of public stakeholders in Chelsea.

There have also been attempts by local residents, who are part of the NoChelseaCrossrail public stakeholder group, to lobby London mayoral candidates, like Zak Goldsmith and Sadiq Khan (Lobo and Abid, 2020), to oppose the plans to build the new Crossrail 2 station in Chelsea. NoChelseaCrossrail has also cast doubts on social media about the local council, the Royal Borough of Kensington & Chelsea's proposed sites for social housing and the opposing group, Chelsea4Crossrail2's intentions and identity (Lobo and Abid, 2020).

While an individual from the other group, Chelsea4Crossrail2, supporting construction of the new station, complained on social media about the long, slow car journey to work in London's west end. This was a personal travel gripe rather than a collective one.

Overall, the lessons learned by the NoCrossrailChelsea group from their campaign against Crossrail2, for future UK infrastructure megaprojects, is to get involved as early as possible. Further, the consultation seemed to be merely a tick-boxing exercise, with public stakeholders' concerns perceived to be ignored. Further, requested information was hard to obtain and, so, they had to use the Freedom of Information Act to obtain information regarding the results of the consultation, for instance. There was a complete lack of trust in the focal firm and in the consultation process, as a result.

The lack of fairness and trust perceived in the stakeholder-firm relationship resulted in specific actions taken by a public stakeholder and its intragroups (Crane, 2018; Hayibor and Collins, 2016) to influence decision making, which, with social media, has exponential reach and impact.

Despite their claim legitimacy, NoCrossrailChelsea felt ignored by the focal firm on Crossrail 2 (Lobo and Abid, 2020). This legitimacy was gained through petitions and responses to the consultation exercise, which showed that the majority were opposed to the new Crossrail 2 station in King's Road, Chelsea. This seems to be the case for the external stakeholders of Jones Hill Wood too who garner support via petitions, legal challenges, and organized walks and protests.

The next section analyzes tweets posted by various external stakeholders opposed to HS2 in Jones Hill Wood. The paper concludes with the contribution to the field, along with the limitations of the study and directions for future research.

6.4 HS2's External Stakeholders

On the gov.uk site for protecting ancient woodlands, ancient trees and veteran trees from development, Forestry Commission and Natural England guidance (known as 'standing advice') states that 'ancient woodland takes hundreds of years to establish and is defined as an irreplaceable habitat. It's important for its wildlife (which include rare and threatened species), soils, recreational value and cultural, historical and landscape value.' There is also advice on mitigation measures, use of buffer zones and compensation measures, like planting new native woodland, alongside other measures, but the latter does not directly replace lost or damaged trees or woodland.

The analysis of posts on Twitter regarding the felling of trees in Jones Hill Wood by HS2, shows that there are myriad conservation and wildlife organizations, protest groups, and individuals who have voiced concerns regarding what they perceive as 'risk [of] undermining every environmental commitment they [government] have ever made about

HS2. @HS2ltd@NaturalEngland' (RSPB England, 15 April 2021). The opposing groups use titles for their organizations that send a strong message of protest to the focal firm, HS2 (Ninan and Sergeeva, 2021) including StopHS2 and HS2 Rebellion.

The sub-sections that follow discuss the views of some of the main external stakeholders around a particular section, Jones Hill Wood, which is in the path of HS2, and HS2's response to them.

6.4.1 Woodland Trust

The Woodland Trust is the UK's largest woodland conservation charity, established since 1972. Despite the opposition of organizations, like the Woodland Trust, with 500,000 members, their claims are ignored, with HS2 stating that it would be planting trees to replace the ones destroyed.

> After @NaturalEngland confirmed to us that #HS2 does not have any licences to disturb bats or damage roosts at Jones' Hill Wood, we're calling for an immediate pause on work at the wood to allow the matter to be fully investigated .woodlandtru.st/OGMzh (*Source*: Twitter, 16 November 2020).

6.4.2 RSPB (Royal Society for the Protection of Birds) England

RSPB first got involved in HS2 as a co-signatory of the Right Lines charter in 2011. RSPB assessed the likely damage of HS2 in response to the first plans, highlighting potential damage to Sites of Special Scientific Interest (SSSIs) and bat Species Protection Plans (SPPS). In 2014, RSPB petitioned against the Hybrid Bill to safeguard this assessment. The petition was withdrawn after numerous assurances from HS2. However, in 2018, when involved in details around mitigation for barn owls, HS2Ltd's inability to produce an effective plan led RSPB to withdraw from the process. Following this, RSPB took it up with the Department for Transport, but no concessions were made. Since then, RSPB has continued to be critical as part of the Re-ThinkHS2coalition (*Source*: Twitter, 15 April 2021).
RSPB's frustration is evidenced in the tweet below:

> HS2Ltd has had many years to plan and deliver this scheme, to carry out standard surveys, and obtain the appropriate licences to carry out its work and safeguard protected species. We should not be having professional ecologists feeling the need to legally challenge the validity of licenses issued by Natural England hours in advance of felling of tree—Tony Whitehead, RSPB England, on Twitter, 15 April 2021.

While an individual, Angela Thompson, in replying to RSPB England, stated on Twitter, 15 April 2021:

> Are the RSPB commenting on the @eastwestrail consultation that is out now? Likely to be significant impacts on rural areas, flood meadows, and farmland thus on birds that use these areas, e.g., skylarks, turtle doves, ravens which we have near us, etc. @CambApproaches.

There was no response to Angelas's question.

6.4.3 StopHS2

Sharing a *Times* article titled 'Roald Dahl's Fantastic Mr. Fox Wood Felled for HS2', on 25 March 2021, Penny Gaines, chairwoman of StopHS2, tweeted:

> The destruction by HS2 Ltd of Jones' Hill Wood is heartbreaking. The UK is one of the least wooded countries in Europe, so even small woodlands matter, especially if they are ancient woodland.

StopHS2 (@stophs2) tweeted on Sunday, August 29, 2021:

> #HS2—started as an incoherent decision and it has been incoherent in its execution as well...If ministers had been told that £100bn would not have bought a complete railway, it might never have gone ahead—Former head of National Audit Office Lord Amyas Morse (FT, 26.8.21) https://t.co/k793MvffEz (https://twitter.com/stophs2/status/1431909265699184643?s=03).

6.4.4 HS2Rebellion

More recently, @HS2Rebellion tweeted on 14 September 2021:

> Campaigners concerned about how driving two 16 km rail tunnels through the Chilterns will affect the area's precious chalk streams have begun legal action against the Environment Agency (EA). endsreport.com/article/1727159via@theENDSReport.

6.4.5 Individuals

There are several individuals, including activist protesters and people living nearby or with a connection to the woods, and politicians who have been very vocal about the felling of Jones Hill Wood.

Pigeon–Pigeon, as she calls herself, is a protester in her mid-20s. Speaking from 9 m high in the branches of a beech tree in Jones Hill Wood, she is 'here to fight for the woodland—I'm making a stand' (bbc.co.uk, 8 April 2021). She was in an hour-long aerial standoff with a bailiff from the National Eviction Team. She has been at the site on-and-off for nearly a year and believes that extreme damage is being done to facilitate building HS2. The latter is deemed an expensive travel option by Pigeon, compared to using a car and 'won't be carbon neutral in its 120-year lifetime.' Three people including Pigeon were arrested, charged with obstructing a public highway, while a fourth was arrested on suspicion of a public order offence dating to March 2021, and were released under investigation (bbc.co.uk, 8 April 2021).

Farmer Bunce—In Roald Dahl's *Fantastic Mr. Fox* (written in 1970), the wily Mr. Fox outwits farmers Bunce, Boggis, and Bean. There is a real-life Bunce family which still farms in the area. Peter Bunce said that 'he played in the wood with his brother and sisters as a child and he was 'gutted' at its destruction.' He said that it used to be 'full of primroses, foxgloves, bluebells, wood anemones, all sorts of things...it's disgusting what they have done' (bbc.co.uk, 8 April 2021).

Mark Keir (MP for Green Party)—@MarkKeir6 tweeted on 30 April 2021 'A last goodbye to Jones Hill Roald Dahl's wood Wendover. Chainsaws and tears.' @MarkKeir6 tweeted on 14 February 2021 'Jones Hill Crew have worked non-stop all winter. The eyes, ears, and voice of so many people and all nature, we have built a legal case to stop work in the wood, but need support. The legal costs £35k. Tomorrow we launch our Crowdjustice fundraiser. Please donate and share.'

A response from a member of the public, SC, below, to the above tweet from Mark Keir on 15 February 2021: 'Replying to @MarkKeir6: Will donate Mark. My mum played in the woods during WW2 child evacuation. Mr. Bunce allowed our family to stay on their farm in a caravan. And now their fields have been TAKEN from them. Sign me up.' Childhood memories have been annihilated along with this ancient woodland.

6.4.6 Focal Firm – HS2 Ltd

HS2 claims to be one of the most sustainable high-speed railways in the world, supporting the UK's transition to a net zero economy. It has committed to plant up to 7 million trees and shrubs along the first phase of the line and create more than 33 sq km (12.7 sq miles) of new woodland, wildlife, and river habitats. As for the remains of Jones' Hill Wood, HS2 plans to create a new woodland, planting 22,000 trees which will link to the remains of Jones Hill Wood, connecting it to another wood. They will 'look at the native species that are already within the landscape and design a mitigation scheme that is sympathetic to what is around it' (Alex Towse, Construction Director with HS2 contractor, Fusion JV; bbc.co.uk, 8 April 2021).

Yet, a petition to scrap HS2, signed by 155,000 people was debated in Parliament on 13 September 2021. It was read into Hansard as:

> 'We ask Parliament to repeal the High Speed Rail Bills, 2016 and 2019, as MPs voted on misleading environmental, financial, and timetable information provided by the Dept. of Transport and HS2 Ltd. It fails to address the conditions of the Paris Accord and costs have risen from £56bn to over £100bn' (*Source*: hansard.parliament.uk, 13 September 2021). The minister agreed to answer in writing all the questions that there was no time to address in the debate.

6.5 Discussion

The words used by the focal firm, HS2, to respond to external stakeholder concerns are non-emotive and concrete; around 'commitment (to plant up to 7 million trees and shrubs), plans (for new woodland) and mitigation (scheme)'; aligned with guidance set out by the Forestry Commission and Natural England. Those who oppose HS2 use words and phrases, like 'gutted, destruction, disgusting, significant impact, heart-breaking, critical, won't be carbon-neutral' to express their frustrations and emotions while they use available recourse like a 'petition, legal challenge, crowdfunding and a coalition' to mobilize and fight for their cause, which is the 'voiceless' natural environment.

The analysis suggests that social media provides emotional support to groups, with social media playing a key role to cohere, empower, and mobilize stakeholders. Though a 'partnership' view was recommended by Lobo and Abid (2020), there is tension and conflict between the external stakeholders and the focal firm on HS2, with arrests of protesters, with the latest petition to stop HS2 resulting in a debate in Parliament on 13 September 2021. To honor the COP26 draft agreement (bbc.co.uk, 12 November 2021), the UK's commitment to a net zero economy by 2050, should be evidenced on all major projects.

6.6 Conclusion

This chapter sought to focus on the 'voiceless' external stakeholder, the natural environment, on infrastructure megaprojects. This external stakeholder has to rely on activists and environmental and conservation organizations to speak on its behalf.

The following are some suggestions, for future research on the role of social media for external stakeholder management. The issues that concern external stakeholders can be discerned by examining posts on social media and used to create a framework for project managers to better manage interactions with and responses to external stakeholders. The examination of social media communication regarding HS2 at Jones' Hill Wood indicates the following: that the focal firm and the external stakeholder discourse are in opposition; the former non-emotive announcing promises, commitments and plans while the latter expresses deep sentiment and emotion including referring to it as 'wildlife crime' (*Source*: *The Guardian*, 10 October 2020), heightened by an embedded lack of trust towards the focal firm, HS2. This may be historical and related to actions taken by HS2, for example in Crackley Wood in Warwickshire, one of the first ancient woods to suffer, despite campaigners' vociferous protests. Discourse analysis is a method that would be very useful to understand the social media content in context, and perhaps to encourage use of posts in a more sympathetic and strategic manner, by project managers.

Since what encourages trust formation in the stakeholder-focal firm relationship is less investigated, this would be a useful area of research and in a particular context. Factors like high quality of information found to be important in WeChat communications in a China context (Cheng et al., 2017) can be inferred to play a part in the infrastructure project context. Perhaps, the Trust Onion Cognitive model developed by Cheng et al. (2017) could be examined and adapted for trust formation in the megaproject environment.

Finally, the natural environment is a 'voiceless' external stakeholder important to the biodiversity of the planet but has to rely on the voice of human external stakeholders, whether individually or in organizations, societies, and coalitions to speak on its behalf. How sustainable is a high-speed railway that may reduce journey times and take cars off the roads but destroy habitats of rare species? The measure for what is a sustainable infrastructure project should be re-examined.

One of the limitations of this study is that it considers only UK infrastructure megaprojects and the social media posts regarding it from external stakeholders and with responses from the focal firm. Future research can explore comparing social media use by external stakeholders of several megaprojects and how the conversations evolve through a longitudinal study of these mega-projects. The use of software for analysis of social media would be a way forward. Future research can also explore and compare international projects' use of social media by the focal firm to respond to and manage external stakeholders.

Acknowledgement

I would like to acknowledge Agha Farhan Abid's contribution as my co-author of the award-winning *Project Management Journal* paper

(Lobo and Abid, 2020), on which this chapter draws to develop the case for future research.

References

Aaltonen, K., Jaakko, K., and Tuomas, O. (2008). Stakeholder salience in global projects, *International Journal of Project Management*, **26**(5): 509–516.

Aaltonen, K., and Sivonen, R. (2009). Response strategies to stakeholder pressures in global projects, *International Journal of Project Management*, **27**(2): 131–141.

Aaltonen, K., and Kujala, J. (2016). Towards an improved understanding of project stakeholder landscapes, *International Journal of Project Management*, **34**(8): 1537–1552.

Altshuler, A.A., and Luberoff, D.E. (2004). *Megaprojects: The Changing Politics of Urban Public Investment*, Washington DC: Brookings Institution Press.

Batheram, M., Hardin, J., and Whitfield, S. (2005). Successful participation methods for local transport planning, *Proceedings of the Institution of Civil Engineers—Municipal Engineer*, **158**(1): 9–16.

Biesenthal, C., Clegg, S., Mahalingam, A., and Sankaran, S. (2018). Applying institutional theories to managing mega projects, *International Journal of Project Management*, **36**(1): 43–54.

Bonke, S., and Winch, G. (2002). Project stakeholder mapping: Analysing the interests of project stakeholders. pp. 385–403. *In:* Slevin, D.P., Cleland, D.I., and Pinto, J.K. (Eds.). *The Frontiers of Project Management Research*. Newtown Square, PA: Project Management Institute.

Bourne, L., and Walker, D.H. (2005). Visualising and mapping stakeholder influence, *Management Decision*, **43**(5): 649–660.

Cheng, X., Fu, S., and de Vreede, G.-J. (2017). Understanding trust influencing factors in social media communication: A qualitative study, *International Journal of In-formation Management*, **37**: 25–35.

Crane, B. (2018). Revisiting who, when, and why stakeholders matter: Trust and stakeholder connectedness, *Business & Society*. doi: 10.1177/0007650318756983.

Daemi, A., Chugh, R., and Kanagarajoo, M.V. (2020). Social media in project management: A systematic narrative literature review, *International Journal of Information Systems and Project Management*, **8**(4): 5–21.

Dekhoda, A., Owens, R.G., and Malpas, P.J. (2020). A Netnographic approach: Views on assisted dying for individuals with dementia, *Qualitative Health Research*, **30**(1): 2077–2091.

Derry, R. (2012). Reclaiming marginalized stakeholders, *Journal of Business Ethics*, 111(2): 253–264.

Desjardins, J. (2017, January 11). 9 of the world's largest megaprojects that are under construction. Retrieved from https://www.businessinsider.com/the-worlds-largest-megaprojects-2017-1?r=US&IR=T.

Dewey, O.F. and Davis, D.E. (2013). Planning, politics, and urban mega-projects in developmental context: Lessons from Mexico City's airport controversy, *Journal of Urban Affairs*, **35**(5): 531–551.

Di Maddaloni, F. and Davis, K. (2018). Project manager's perception of the local communities' stakeholder in megaprojects. An empirical investigation in the UK, *International Journal of Project Management*, **36**: 542–565.

Fewings, P. (2013). *Construction Project Management: An Integrated Approach*, London, England and New York, NY: Routledge.

Flyvbjerg, B. (2014). What you should know about megaprojects and why: An overview, *Project Management Journal*, **45**(2): 6–19.

Freeman, R. (1984). *Stakeholder Management: A Strategic Approach*, New York, NY: Pitman.

Freeman, R.E., Harrison, J., Wicks, A., Parmar, B., and de Colle, S. (2010). *Stakeholder Theory: The State of the Art*, Cambridge, UK: Cambridge University Press.

Frooman, J. (1999). Stakeholder influence strategies, *The Academy of Management Re-view*, **24**(2): 191–205.

Frooman, J., and Murrell, A. (2005). Stakeholder influence strategies: The roles of structural and demographic determinants, *Business and Society*, **44**(1): 3–31.

Germain, J., Harris, J., and Maxwell, M.C. (2018). Why should we use online research methods? Four doctoral health perspectives, *Qualitative Health Research*, **28**(10): 1650–1657.

Giezen, M. (2013). Adaptive and strategic capacity: Navigating megaprojects through uncertainty and complexity, *Environment and Planning B: Urban Analytics and City Science*, **40**(4): 723–741.

Greene, J.A., Choudhry, N.K., Kilabuk, E., and Shrank, W.H. (2011). Online social networking by patients with diabetes: A qualitative evaluation of communication with Facebook, *Journal of General Internal Medicine*, **26**: 287–292.

Hayibor, S., and Collins, C. (2016). Motivators of mobilisation: Influences of inequity, expectancy, and resource dependence on stakeholder propensity to take action against the firm, *Journal of Business Ethics*, **139**(2): 351–374.

Jepsen, A.L., and Eskerod, P. (2009). Stakeholder analysis in projects: Challenges in using current guidelines in the real world, *International Journal of Project Management*, **27**(4): 335 –343.

Kozinets, R.V. (2010). Netnography: Doing ethnographic research online, *Thousand Oaks*, CA: Sage Publications.

Kozinets, R.V., Dolbec, P.Y., and Earley, A. (2014). Netnographic analysis: Understanding culture through social media data. pp. 262–276. *In:* Flick, U. (Ed.). *The SAGE Handbook of Qualitative Data Analysis*, London: Sage.

Lehtinen, J., Aaltonen, K., and Rajala, R. (2019). Stakeholder management in complex product systems: Practices and rationales for engagement and disengagement, *Industrial Marketing Management*, **79**: 58–70.

Lincoln, Y.S., and Guba, E.G. (1985). *Naturalistic Inquiry*, Newbury Park, CA: Sage Publications.

Lobo, S., and Abid, A.F. (2020). The role of social media in intrastakeholder strategies to influence decision making in a UK infrastructure Megaproject: Crossrail 2, *Project Management Journal*, **51**: 96–119.

Ma, G., Jiang, S., and Jia, J. (2021). Investigating the adoption of social media in the construction industry: Eempirical evidence from project teams in China, *Engineering, Construction and Architectural Management*. doi: 10.1108/ECAM-05-2021-0456.

Mathur, S., Ninan, J., Vuorinen, L., Ke, Y., and Sankaran, S. (2021). An exploratory study of the use of social media to assess benefits realisation in transport infrastructure projects in transport infrastructure projects, *Project Leadership and Society*, 2: 100010.

Mok, K.Y., Shen, G.Q., and Yang, J. (2015). Stakeholder management studies in megaconstruction projects: A review and future directions, *International Journal of Project Management*, **33**(2): 446–457.

Mok, K.Y., Shen, G.Q., Yang, J., and Li, C.Z. (2017). Investigating key challenges in major public engineering projects by a network-theory based analysis of stakeholder concerns: A case study, *International Journal of Project Management*, **35**(1): 78–94.

Nikolaidou, A., and Papaioannou, P. (2018). Utilising social media in transport planning and public transit quality: Survey of literature, *J. Transp. Eng., Part A: Systems*, **144**(4): 04018007.

Ninan, J., Mahalingam, A., and Clegg, S. (2019a). External stakeholder management strategies and resources in megaprojects: An organisational power perspective, *Project Management Journal*, **50**(6): 625–640.

Ninan, J., Clegg, S., and Mahalingam, A. (2019b). Branding and governmentality for infra-structure megaprojects: The role of social media, *International Journal of Project Management*, **37**: 59–72.

Ninan, J. (2020). Online naturalistic inquiry in project management research: Directions for research, *Project Leadership and Society*, **1**(1): 1–9.

Ninan, J., Mahalingam, A., and Clegg, S. (2020a). Power and strategies in the external stakeholder management of megaprojects: A circuitry framework, *Engineering Project Organisation Journal*, **9**(1): 1–20.

Ninan, J., Mahalingam, A., Clegg, S., and Sankaran, S. (2020b). ICT for external stakeholder management: Sociomateriality from a power perspective, *Construction Management & Economics*, **38**(9): 840–855.

Ninan, J., and Sergeeva, N. (2021). Labyrinth of labels: Narrative constructions of promoters and protesters in megaprojects, *International Journal of Project Management*, **39**(5): 496–506.

Olander, S., and Landin, A. (2005). Evaluation of stakeholder influence in the implementation of construction projects, *International Journal of Project Management*, **23**(4): 321–328.

Olander, S., and Landin, A. (2008). A comparative study of factors affecting the external stakeholder management process, *Construction Management and Economics*, **26**(6): 553–561.

Oliveira, G.F., and Rabechini, R., Jr. (2019). Stakeholder management influence on trust in a project: A quantitative study, *International Journal of Project Management*, **37**(1): 131–144.

Pirson, M., Martin, K., and Parmar, B. (2016). Public trust in business and its determinants, *Business & Society*, **58**(1): 132–166.

Pirson, M., Martin, K., and Parmar, B. (2017). Formation of stakeholder trust in business and the role of personal values, *Journal of Business Ethics*, **145**(1): 1–20.

Shandraseharan, A., and Kulatunga, U. (2021). Social media for disaster management: the Sri Lankan context, *International Journal of Construction Management*, doi: 10.1080/15623599.2021.1907015.

Suchman, M.C. (1995). Managing legitimacy: Strategic and institutional approaches, *The Academy of Management Review*, **20**(3): 571–610.

Tang, L.Y., Zhang, Y.M., Dai, F.,Yoon, Y., Song, Y., and Sharma, R.S. (2017). Social media data analytics for the U.S. construction industry: Preliminary study on twitter, *J. Manage. Eng.*, **33**(6): 04017038.

Teo, M., and Loosemore, M. (2014). The role of core protest group members in sustaining protest against controversial construction and engineering projects, *Habitat International*, **44**: 41–49.

van den Ende, L., and van Marrewijk, A. (2019). Teargas, taboo and transformation: A neo-institutional study of community resistance and the struggle to legitimize subway projects in Amsterdam 1960–2018, *International Journal of Project Management*, **37**(2): 331–346.

Williams, N.L., Ferdinand, N., and Pasian, B. (2015). Online stakeholder interactions in the early stage of a megaproject, *Project Management Journal*, **46**(6): 92–110.

Yang, J., Shen, G.Q., Ho, M., Drew, D.S., and Xue, X. (2011). Stakeholder management in construction: An empirical study to address research gaps in previous studies, *International Journal of Project Management*, **29**(7): 900–910.

7

Social Media for Improving Metro Rail Project Operations

Johan Ninan,[1,*] *Yongjian Ke,*[2] *Shankar Sankaran,*[2] *Sandeep Mathur,*[3] *Lauri Vuorinen*[4] and *Ganesh Devkar*[5]

7.1 Introduction

Countries across the world are investing heavily in infrastructure projects to achieve their socio-economic objectives. These projects are created with enormous budgets and are often called as infrastructure megaprojects, i.e., projects costing more than one billion USD (Flyvbjerg, 2014). Wang et al. (2020) suggested that the boom in megaprojects is having 'a significant impact on social and economic developments (e.g., immigrant settlement, poverty eradication, and public health) as well as the natural environment (e.g., ecological processes and biodiversity)' (p. 831). Majority of these expenses are incurred in the planning and construction phase, due to which the front-end of these projects draws attention from policy makers, construction companies, designers, and researchers (Osei-Kyei and Chan, 2015). However, the value to the society or impact on the society is realized only in the later periods of these projects, particularly the operation phase and projects are remembered for how they serve the community more

[1] Delft University of Technology (TU Delft), Netherlands.
[2] University of Technology Sydney, Australia.
[3] Transport New South Wales, Australia.
[4] Tampere University, Finland.
[5] CEPT University, India.
* Corresponding author: Johan.Ninan@gmail.com

than the efficiency of construction phase. For example, the Sydney Opera House was a project-management failure, resulting in cost overrun and time overrun but is now an iconic symbol for Sydney and even Australia.

Most infrastructure projects have a lifecycle of 60–80 years, with the initial four to five years earmarked for planning and the next four to five years for construction. The remaining period of the infrastructure project is its operation phase, which takes up the largest portion of the project's lifecycle (Liu et al., 2014). An infrastructure project with a poorly managed operational phase undermines the project objectives and erodes its value for money (Osei-Kyei et al., 2017). In contrast, a high performing infrastructure project is generally associated with high operational efficiency and addressing the needs of the society (Partnerships UK, 2006). The consistent ignoring of the operations phase by the designers, contractors, policy makers, and researchers has led to recurrent occurrence of problems and thereby increased lifecycle costs of these projects (Karim and Magnusson, 2008). Projects require an integration of its life-cycle dynamism, stakeholder heterogeneity, and social responsibility interactivity (Lin et al., 2017), most of which are only evident during the operation phase.

The research reported in this chapter discusses how the operation phase of infrastructure projects can be managed effectively. Frequent monitoring of the operations of a project can help the decision-makers to accurately evaluate the performance objectives and take corrective actions to increase value for the society (Osei-Kyei et al., 2017). Considering project management in the 21st century, inferences to operation management can be drawn from the digital big data. Big data sets are increasingly used to help public managers derive real-time insights into behavioral changes, public opinion, or daily life (Mergel et al., 2016), particularly when we look at the big data available on social media regarding the project. Megaprojects, due to their large scale, complex and controversial nature (Frick et al., 2008), attract a lot of attention on these digital platforms, with users taking it to social media to express their sentiments regarding the project. The large social media data can be mined to address concerns of the community and improve services during the operation phase of infrastructure projects. Social media is one of the significant channels for constructing voice of customer which is increasingly being used to provide qualitative feedback on benefits tracking and data-driven decisioning. Data Science Initiatives (DSI) which include investments in data analytics, business intelligence, and data science including machine learning and Artificial Intelligence (Mathur et al., 2021), provide us with a mechanism to harness the big data including that on social media, generated during delivery and in operation phase. Mathur et al. (2021) proposed a DSI Delivery Framework synthesizing programme management, change management, scaled agile, data management, and data science domains for programme managers to realize value from such investments. Hence, our research objective is to develop a framework for engaging with the users of an infrastructure project during the operation phase, using social media big data.

This chapter is organized as follows. In the next section, the literature on challenges in operation phase, the use of big data, and the role of social media in projects are discussed. Following this, the case study of two metro rail projects in Sydney, Australia and Chennai, India are introduced along with the data collection and analysis methods employed. The findings section discusses how social media can be an effective way to manage the operation phase of infrastructure projects. Subsequently, the challenges of using social media platform and the proposed data science initiative delivery framework are discussed. The final section highlights the contributions, limitations, and scope for future research in the area.

7.2 Literature Review

7.2.1 Importance of Operational Phase for Projects

The operations phase of an infrastructure asset covers the running, repair, and maintenance of the project (Liu et al., 2014). There can be multiple concerns during the operation phase of an infrastructure asset that can affect the long-term demand of the facility (Osei-Kyei et al., 2017). Research highlights that the most reported problems in infrastructure projects are evident during the operational stages, and some are exclusively confined to this phase (Edwards et al., 2004). The application of improper operations and maintenance methods can reduce the productivity of the asset and value for the users (Ghalenoeei et al., 2021). It is also important to have effective safety and environmental health management systems in place for successful operations of an infrastructure asset (Liu et al., 2014).

Despite the importance of the operations phase, maintainability of an infrastructure project is seldom considered a priority for designers and policymakers (Barbarosoglu and Arditi, 2019). Many scholars (Brochner, 2003; Meng, 2013; Wang and Xie, 2002) argue that the problems in the operations phase of an infrastructure asset result from wrong processes and decisions from the early phases of projects. In short, the decisions made in the initial stages of the project can affect the fate of the asset during its lifecycle (Dunston and Williamson, 1999). Others claim the problems arise due to lack of constructive feedbacks from the operations and maintenance staff to the design team during the planning phase (Jensen, 2009). It is appropriate and cost effective to consider various aspects of operations in the early stages of projects and revise the project parameters, considering that the ability to change is high and the cost to change is low during the early stages (Gardiner and Simmons, 1992). Thus, frameworks have been created to bring the operators and maintenance professionals together with designers during the planning phase, to address the future operational issues (Rivera-Vazquez et al., 2009). Even the use of personnel knowledge and experience throughout the life cycle are considered by designers for creating a project well suited to operation phase (Liu and Issa, 2016).

7.2.2 Foreseeing Operational Issues

It is rarely possible for projects to foresee all the operational issues in the front-end itself, especially considering the long life of these projects. In some cases, knowledge of operations in different societies are contradictory, making its translation to design stages complex (Cheng, 2004). Even within the same context, it is not possible for planners to foresee all the challenges that the project would experience during its lifetime. As Mok et al. (2015) highlight, project managers are unable to comprehend the full project lifecycle because of their limited cognition. Unforeseen events can occur in projects during its lifecycle as it is not possible for even experts to forecast anything over a year (Flyvbjerg, 2021). Ma et al. (2017) argued that 'the prolonged lifecycle and heterogeneous stakeholders of megaprojects have posed great challenges for the governance of the economic, social, and environmental issues involved' (p. 1365). Even though the ability to make changes in a project decreases with time, some adjustments can still be made to improve the experience of the users. So, it is important to listen to the users of the infrastructure asset and take necessary actions for improving operations.

Reliable and effective service delivery is important as any public project aims to achieve service as the end result (Robinson and Scott, 2009; Meng et al., 2011). The service delivery framework of project needs to be periodically evaluated and updated, based on the changing needs of the society (Partnerships UK, 2006; Mladenovic et al., 2013). Scholars have also suggested that user-fee adjustments have to be made routinely and stakeholders including public users and trade unions have to be consulted prior to these adjustments (Kumaraswamy and Morris, 2002; Edwards et al., 2004). Additionally, there should be a focus on timely rectification of operational problems encountered by the users (Partnerships UK, 2006). For all these, there should be proper stakeholder engagement not just during the front end of a project, but also during the operation phase. As noted by Tang et al. (2012), transparency and frequent communication with external stakeholders during the operation phase are critical factors in the success of infrastructure projects. There is a need to create an efficient system for bringing in feedbacks from the users to improve the infrastructure service and manage the dynamic operation phases of infrastructure assets.

7.2.3 Stakeholder Engagement in the Backend of Projects

Stakeholder engagement has traditionally been restricted to the front end of infrastructure projects. For example, stakeholder engagement in co-creation sessions helped in co-creation of public values in front end of infrastructure development programmes (Liu et al., 2019; Oldenhof et al., 2014; Steenhuisen and van Eeten, 2008). However, there is a growing argument that stakeholder engagement should not be restricted to just the planning stages of a megaproject, but also extend to the construction and operation phase (Ninan et al., 2020). Stakeholders have to be engaged during the operation phase to considerably inform practitioners of the key operational management issues which need critical attention (Osei-Kyei et al., 2017). Traditionally these are done through surveys, interviews, questionnaires, and feedback forms.

In the modern era, data from different online sources can be compiled to manage projects better (Ninan, 2020). Big data is the large, diverse, complex, longitudinal, and/or distributed datasets generated from instruments, sensors, internet transactions, e-mail, video, click streams, and/or all other digital sources available today and in the future (Favaretto et al., 2020; p.3). Big datasets are increasingly used to help public managers derive real-time insights into behavioral changes, public opinion, or daily life (Mergel et al., 2016). Big data has been used to evaluate customer agility and responsiveness for public value creation. It is predicted that new smart technologies and strategies will shape and will be shaped by the future of public organizations and management and could lead to transformative practices in the public sector. There are already some examples from around the world that big data has been useful in providing more transparent information to the public and citizens, such as healthcare (Raghupathi and Raghupathi, 2014; Ziora, 2015); public sector performance (Bovaird and Loeffler, 2015), and smart cities (Al Nuaimi et al., 2015).

Big data can be generated from a wide range of sources to create value in infrastructure megaprojects. Sensors can record the people's movement in stations (Alawad and Kaewunruen, 2018) and help decision-makers increase or decrease the frequency of trains for certain periods of the day. The way people move within the stations can also help decision-makers allocate commercial space in the stations effectively. Another source of big data generated by infrastructure users is the data available on social media. Social media data can enable us to understand megaprojects better as many conversations relating to projects are only evident online and they are not currently captured or analyzed (Ninan, 2020). Even though there are large amounts of data generated in the social media regarding infrastructure projects daily, researchers have not explored the scope of using this big data for decision-making in infrastructure projects. Social media has been used in projects for studying branding practices (Ninan et al., 2019), requirements management (Daemi et al., 2020), consultation process (Lobo and Abid, 2020), stakeholder engagement process (Ninan et al., 2020; Lehtinen, 2021), and knowledge sharing (Ma et al., 2021). It is in this context that we situate our research to understand how social media can be used to manage the operation phase in infrastructure megaprojects. In one of our earlier research projects, we found that there were more tweets with complaints about the infrastructure service than compliments about it (Mathur et al., 2021). People generally log into social media to criticize rather than to praise (Park, 2015). As noted by Dwivedi et al. (2021), negative information flows faster than positive information in social media platforms. Project operators can mine this data source for a good understanding of the concerns of the community during the operation phase and then address these to improve services.

7.3 Research Setting and Method

To understand how the big data from social media can be used to manage the megaproject operations, we chose a qualitative research methodology. Scholars have suggested that such a method is apt for exploratory research when the aim is to gain familiarity with a problem or to generate new insights for future research (Eisenhardt, 1989; Scott, 1965). Within the qualitative research methodologies, we chose to use the multiple case study method as it provides excellent opportunities to enhance contextual understanding and simultaneously enable the generalization of findings (Flyvbjerg, 2006; Saunders et al., 2009). Multiple-case design brings benefits of robustness, versatility, and replication (Yin, 2017).

We chose the case study of a metro rail project in Australia and India to understand the scope of using social media for managing the operations phase. We selected the Sydney metro rail project from Australia and the Chennai metro rail project from India because of their similarities. Both the projects started operation around the same time and had similar track length: 36 kms and 45.1 kms respectively. Twitter posts of the Sydney metro rail project and Chennai metro rail project were captured through a Twitter search API. The keywords are the titles of two projects, i.e., 'Chennai Metro' and 'Sydney Metro'. It is acknowledged that some tweets would not be retrieved if those tweets discussed the two projects without using the keywords. No duplicates were observed on checking the unique ID of each tweet, and the collected data were stored as a comma-separated values (CSV) file. We collected the tweets for a 90-day period from 1 July

Table 7.1: Summary of Cases Considered.

	Sydney Metro Rail	**Chennai Metro Rail**
Operation start date	26 May 2019	10 February 2019
Operational network length	36 kms	45.1 kms
Keyword used for search	'Sydney Metro'	'Chennai Metro'
Period of data collected	1 July 2019 to 30 September 2019	1 July 2019 to 30 September 2019
Tweets collected	5960	1064

2019 to 30 September 2019, during which both the metro rail projects were operational. There were 5,960 tweets relating to the Sydney metro rail project and 1,064 tweets relating to the Chennai metro rail project. All the tweets were in English.

For analysis, we used content analysis and open coding of the tweets collected to understand what each tweet conveyed. The process was very iterative, and we took multiple readings of the tweets as some categories are often not obvious until the second or third reading (Steger, 2007), due to the focus on content and meaning. We employed manual coding as automatic methods could create a barrier to our understanding in this exploratory study (Kozinets et al., 2014).

7.4 Findings

From the analysis of the social media tweets collected from the Sydney metro rail and Chennai metro rail, we found that social media can be used for collating feedbacks from the users and charting the future direction of the infrastructure project. Social media can inform project operators on the day-to-day technical operational issues, collating the suggestions to improve, and capturing the live sentiments associated with the project.

7.4.1 Social Media for Addressing Day-to-day Technical Operational Issues

Issues relating to operation of the infrastructure service have to be addressed as soon as possible for smooth service. The users widely shared operational issues relating to the project on Twitter, across both projects. In Chennai metro rail, one user complained in Twitter that the doors were not opening in one of the stations as below:

> 'Crazy. @chennaimetro rail's doors didn't open when it stopped @Pachaiyappas metro station, at around 11 am today, putting the passengers to hardships. What's happening?' (2 Sept. 2019)

Similarly, in the case of Sydney metro rail, a user complained about lifts being out of service in one of the stations.

> 'The lift between the concourse and the platforms at North Ryde is out of service' (29 Sept. 2019)

In the case of the Sydney metro rail project, besides complaints about specific facilities, such as USB points, air conditioning, thermometers and escalators, the most common complaints were about train delays. Social media provides an excellent platform where users of the infrastructure service post day-to-day operational issues surrounding the projects. We can create more value in infrastructure projects if we can systematically collect this big data, analyze them through algorithms, and efficiently communicate it to the service team to mitigate the current issue through timely action.

7.4.2 Social Media for Collating Suggestions to Improve

Many users are active stakeholders offering multiple suggestions to improve the services. In contrast to operational issues, suggestions to improve are more than addressing an operational defect on a particular day. For example, in the case of the Sydney metro, a user offered a suggestion to fix the 15-second door-opening duration before people get hurt, as below:

> '@SydneyMetro Your 15-second door opening is stupid and dangerous. People cannot get off the train in the fifteen seconds. Fix it before people get hurt' (16 Sept. 2019).

Similarly, in the case of Chennai metro, a user suggested to bring down the ticket costs and this will lead to more traffic and hence revenue, as below:

> 'Volumes should be the mantra & increased patronage will automatically bring in more revenue & help in bridging gap between cost & income.' (30 Sept. 2019).

In the case of the Chennai metro rail project, the community complained about different aspects of the project, such as its unaffordability and poor design. Collating such suggestions to improve can help the project create more value for the society as decision makers would know the main issues raised by the community.

7.4.3 Social Media for Capturing the Live Sentiments Associated with the Project

The users also communicated their feelings regarding the project on social media. They posted selfies and took pride in the megaproject being in their city. In one instance, a user in the Sydney metro rail claimed he was having fun and felt like he was on a holiday, as below:

> 'Tbh going on the Sydney metro makes me feel like I'm on holidays … So fun' (30 Sept. 2019)

In the case of Chennai metro rail, a user claimed that the project is a step towards public transport as many are leaving their cars and choosing to travel by metro. The user also claims that the metro rail is a safe, convenient, and clean means of transport, as highlighted below:

> 'Yes, of course. It's getting there. It's visible in office as there are many of us who leave our cars at the station and take the Metro. So many people exercising this option that safe and convenient and clean' (20 Aug. 2019)

Capturing live sentiments can give guidance to investors to make long-term strategic decisions, such as which route has to be expanded. It can also help the project take suitable steps to improve the sentiment associated with the project, such as offering complimentary rides for school children or celebrating a regional festival (Ninan et al., 2019). By using big data from social media to analyze sentiments, decision-makers can take proper strategic decisions to create more value to the public.

7.5 Discussion

As discussed above, social media can help manage the operation phase of infrastructure projects; however there are certain challenges in analyzing the data. After discussing these challenges, we will highlight the steps to create a data science initiative delivery framework for using social media in operations management of infrastructure megaprojects.

7.5.1 Challenges of using the Social Media Platform

The major challenge to deriving inferences from big data is the bias in data. The presence of interest groups and predominance of younger group can bias the findings from social media.

1. *Presence of interest groups*: One of the pressing issues of public infrastructure projects is a change for the most vocal, instead of most affected. The most vocal opposition tend to challenge infrastructure development and often get what they want from the project (Bornstein, 2010). In the case of the Chennai metro rail project, there was an interest group campaigning against land acquisition, often posting the same message daily on social media. Similarly, some tweets were often re-tweeted, and this can also create a challenge in analyzing the big

data and making decisions relating to suggestions to improve the service.

2. *Predominance of younger groups*: One of the most highlighted limitations of social media data is its inability to represent the entire population. Younger population have substantial technical knowledge, digital familiarity, and willingness to engage online and, hence, are more represented on social media (Kaplan and Haenlein, 2010), which can bias the data. Critical insights for the infrastructure operation, such as elderly care and access ramps may not be represented in social media discourses. Therefore, the insights gained from the analysis on social media on the way people behave may not be representative of all age groups and needs to be coupled with traditional feedback mechanisms, such as feedback forms or suggestions from customer service counters.

To address these challenges, a data science initiative framework, which can reduce bias due to presence of interest groups and predominance of younger groups needs to be implemented.

7.5.2 Data Science Initiative (DSI) Delivery Framework

Data science initiative is 'a set of fundamental principles that support and guide the principled extraction of information and knowledge from data' (Provost and Fawcett, 2013). DSIs have unique characteristics and pose challenges in delivering the envisaged value when using traditional processes for managing ICT-enabled programmes (Mathur et al., 2021). Infrastructure assets, such as metro rails, bridges, dykes, and roads are by nature data-intensive and data-science initiatives are required for the digital transformation and management of these assets (Brous et al., 2020; Berman, 2012). Organizations without data-science capability may find it difficult to compete in a technology-intensive environment. Sensing this, many infrastructure asset managers are introducing data-science initiatives to support predictive maintenance modelling and anomaly detection (Chen et al., 2012; van der Aalst, 2016). Our proposed framework seeks to use social media data for predictive analysis and would help in the operations management of infrastructure assets.

Data-science initiatives are complex and encompass a broad range of knowledge and capabilities, such as data mining and machine learning, which are designed to extract knowledge from data (Brous et al., 2020). The quality of the data being used determines the decision-outcomes of data science models and thereby the success of the initiative (Lee and Pan, 2017). A data governance framework defined as the exercise of authority and control over the management of data assets (DAMA International, 2017) can mitigate some of the challenges associated with the initiative and provide benefits. A 'one size fits all' approach may not work for these initiatives as all of them come with their unique set of challenges and the initiative can change, depending on the domains and organizational types (Abraham et al., 2019). For the data-science initiative to manage the operations phase of metro rail projects, the following steps are proposed:

1. *Raw data*: As discussed above, the data from Twitter can be used to manage the operations phase of metro rail projects by informing the day-to-day technical challenges, collating suggestions to improve, and capturing the live sentiments associated with the project. Multiple data retrieval programmes are available which can help retrieve the raw data from these platforms. Suitable keywords can be identified to retrieve social media messages for a particular project.
2. *Process data*: The raw data has to be pre-processed to make it suitable for analysis. This includes removing identifiers to anonymize the data. Automatic data collection may cause privacy infringements, both in the physical space and digital space. For example, cameras in the physical space used to track traffic movements need to remove personally identifiable data, such as number plates or faces of persons in the vehicles (Cecere et al., 2015). Similarly, social media exchanges in the digital space used to manage operations phase of infrastructure projects needs to remove personally identifiable data, such as names and organizations (Ninan, 2020). In addition to pre-processing to remove identifiers, noise, such as use of symbols and stopwords like 'a' and 'the' needs to be removed. Lemmatization needs be carried out to convert multiple words in tweets meaning the same thing, such as drive, drove, driven into a common word. Tokenization is a final step of pre-processing where a particular text is converted to a number to enable analysis and interpretation.
3. *Evaluate data*: There are multiple Natural Language Processing (NLP) and machine learning (ML) models that can analyze the tweets to derive insights. Choosing a particular model is often an iterative process as all models may not apply to a particular context. Analysis of big data can be descriptive, predictive or prescriptive and thereby can inform different decisions during the operation phase.
4. *Interpret data*: Data can be interpreted through Aartificial Intelligence to prescribe what could be done to address operational issues. A distinction can be made between the changes undertaken. Quick fixes can be proposed for addressing day-to-day technical issues. A strategic long-term implementation would be required for addressing the suggestions to improve or steps that could be taken to maintain the sentiments.

The interpreted data has to be integrated into the business process. The four processes described above operate in an iterative process for validation and refining, as shown in Fig. 7.1.

Data science initiatives are delivered through programme management and involves a continuous improvement through a Plan-Do-Check-Act (PDCA) cycle. During the initial programme management phase, only the owner and director of operations may be involved. Change management has to be carried out such that all stakeholders are able to adopt the initiative. Awareness and training programmes can be carried out for the users on how to use social media to reach the project and for the operators on how to implement the insights generated from social media. A change manager may be required for this phase along with a business analyst. Initially, only a small scope is taken up through the DSI initiative. Subsequent scaling of the DSI framework is taken up, based on the feedback from earlier iterations. Agile frameworks can be implemented for scaling the DSI implementation. Support can be taken from agile coaches, developers, project managers, and testers during this phase. Data management is an important step in the DSI framework. The collected data needs to be stored safely after removing the identifiers from the data. For the data-science stage, data scientists, analysts, project manager, and even the director of operations may be involved. The DSI framework for social media used during operations phase is summarized in Table 7.2.

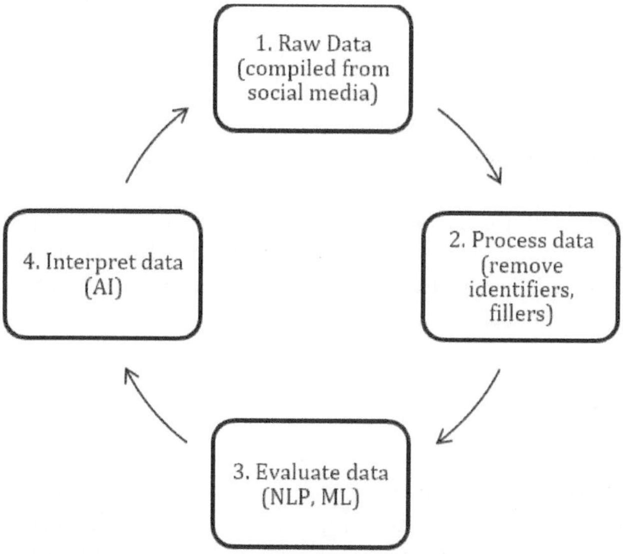

Fig. 7.1: DSI process for using social media to improve metro rail operations.
(Alt Text: A cyclic DSI process framework comprising of four steps – raw data, process data, evaluate data, and interpret data)

Table 7.2: DSI Delivery Framework.

Domains	Key Actions	People
Program Management	PDCA cyclic process	Owner, Director (Operations)
Change management	Awareness creation; training programmes; working concept	Change manager, business analyst, employees and customers
Scaled agile	Start small and scale (one or two action points initially)	Agile coach, developers, project managers, testers
Data management	Data storage and retrieval; data security; automated pre-processing	Data manager, data scientist, data analyst
Data science	Prescriptive and predictive analysis; model validation and implementation	Data scientist, data analyst, project manager, director (operations)

7.5.3 Social Media for Active Transport at Transport for NSW, Australia

While our original study focused on Sydney Metro and Chennai Metro, further use case emerged in active transport mode at Transport for NSW, Australia. Active transport focuses on increasing walking and cycling usage in NSW and provides a healthy and safe option to commute, especially in the post COVID-19 world. Transport for NSW has allocated $950 million to deliver this infrastructure in the state, covering both Greater Sydney and regional and outer metropolitan areas. The key challenge is how to track the benefits of infrastructure which has been delivered and secondly, how to plan connected cycleways and walkways. Trials are underway using cameras to track the pedestrian and cyclist volumes as well as to get additional data on near misses and incidents. This quantitative view on walking and cycling is being complemented with a qualitative view of voice of customer using social media and surveys. A stakeholder and commuter sentiment score is being constructed to assess impact of active transport infrastructure on a near-real time basis. The advances in big-data technology are allowing Transport for NSW network insights to provide better service to the customers.

7.6 Conclusion

This research calls for focus on the operations phase of the project and the use of big data to generate understanding of projects. This exploratory research is aimed at understanding how social media can be used for managing operations in infrastructure projects. Compiling insights from social media can present a comprehensive view of the interests of all stakeholders throughout an infrastructure project's life cycle, as both external and internal stakeholders use social media and are part of the broader project community (Ninan et al., 2021).

The different data on the performance of an infrastructure asset available from social media, such as day-to-day technical issues, suggestions to improve, and live sentiments are discussed from the case study of the Sydney metro rail and the Chennai metro rail projects. Some of the challenges of using the social media platform for operations phase are discussed along with a framework for adoption of the data-science initiative in projects. The use of social media for analyzing feedbacks and engaging with the user comments affords multiple benefits, such as offering real-time data, identifying the major concern of the users, and presenting an intuitive tool for engaging with the community. Such a DSI initiative can pave the way for managers and stakeholders involved in projects to reduce problems, duplications, costs and to increase success of infrastructure projects in their operations. In the future, action research or autoethnography (Datta et al., 2020) methodologies can be used to understand how projects implement the proposed framework, the challenges associated with implementation, and lessons learnt.

References

Abraham, R., Schneider, J., and Vom Brocke, J. (2019). Data governance: A conceptual framework, structured review, and research agenda, *International Journal of Information Management*, **49**: 424–438.

Al Nuaimi, E., Al Neyadi, H., Mohamed, N., and Al-Jaroodi, J. (2015). Applications of big data to smart cities, *Journal of Internet Services and Applications*, **6**(1): 1–15.

Alawad, H., and Kaewunruen, S. (2018). Wireless sensor networks: Toward smarter railway stations, *Infrastructures*, **3**(3): 24.

Barbarosoglu, B.V., and Arditi, D. (2019). A system for early detection of maintainability issues using BIM. pp. 335–341. *In*: Mutis, I. and Hartmann, T. (Eds.). *Advances in Informatics and Computing in Civil and Construction Engineering*, Springer, Cham.

Berman, S.J. (2012). Digital transformation: Opportunities to create new business models, *Strategy & Leadership*, **40**(2): 16–24.

Bornstein, L. (2010). Mega-projects, city-building and community benefits, *City, Culture and Society*, **1**(4): 199–206.

Bovaird, T., and Loeffler, E. (2015). *Public Management and Governance*, Routledge, Abingdon.

Bröchner, J. (2003). Integrated development of facilities design and services, *Journal of Performance of Constructed Facilities*, **17**(1): 19–23.

Brous, P., Janssen, M., and Krans, R. (2020). Data governance as success factor for data science, *Responsible Design, Implementation and Use of Information and Communication Technology*, 12066: 431.

Cecere, G., Le Guel, F., and Soulié, N. (2015). Perceived internet privacy concerns on social networks in Europe, *Technological Forecasting and Social Change*, **96**: 277–287.

Chen, H. and Chiang, R.H., and Storey, V.C. (2012). Business intelligence and analytics: From big data to big impact, *MIS Quarterly*, 1165–1188.

Cheng, Y.C. (2005). Fostering local knowledge and human development in globalization of education, *New Paradigm for Re-engineering Education: Globalization, Localization and Individualization*, 73–94.

Datta, A., Ninan, J., and Sankaran, S. (2020). 4D visualisation to bridge the knowing-doing gap in megaprojects: An Australian case study, *Construction Economics and Building*, **20**(4): 25–41.

DAMA International (2017). *DAMA-DMBOK: Data Management Body of Knowledge*, Technics Publications.

Daemi, A., Chugh, R., and Kanagarajoo, M.V. (2021). Social media in project management: A systematic narrative literature review, *International Journal of Information Systems and Project Management*, **8**(4): 5–21.

Dunston, P.S., and Williamson, C.E. (1999). Incorporating maintainability in constructability review process, *Journal of Management in Engineering*, **15**(5): 56–60.

Dwivedi, Y.K., Ismagilova, E., Hughes, D.L., Carlson, J., Filieri, R., Jacobson, J., Jain, V., Karjaluoto, H., Kefi, H., Krishen, A.S., Kumar, V., Rahman, M.M., Raman, R., Rauschnabel, P.A., Rowley, J., Salo, J., Tran, G.A., and Wang, Y. (2021). Setting the future of digital and social media marketing research: Perspectives and research propositions, *International Journal of Information Management*, **59**: 102168.

Edwards, P., Shaoul, J., Stafford, A., and Arblaster, L. (2004). *Evaluating the Operation of PFI in Roads and Hospitals* (vol. 84), London: Association of Chartered Certified Accountants.

Eisenhardt, K.M. (1989). Building theories from case study research, *Academy of Management Review*, **14**(4): 532–550.

Favaretto, M., De Clercq, E., Schneble, C.O., and Elger, B.S. (2020). What is your definition of Big Data? Researchers' understanding of the phenomenon of the decade, *PloS One*, **15**(2): e0228987.

Flyvbjerg, B. (2006). Five misunderstandings about case-study research, *Qualitative Inquiry*, **12**(2): 219–245.

Flyvbjerg, B. (2014). What you should know about megaprojects and why: An overview, *Project Management Journal*, **45**(2): 6–19.

Flyvbjerg, B. (2021). Make Megaprojects More Modular, *Harvard Business Review*, **6**: 58–63.

Frick, K.T. (2008). The cost of the technological sublime: Daring ingenuity and the new San Francisco–Oakland Bay Bridge. pp. 239–262. *In:* Priemus, H., Flyvbjerg, B., and van Wee, B. (Eds.). *Decision-making on Mega Projects: Cost–benefit Analysis, Planning, and Innovation*, Cheltenham, UK, Edward Elgar.

Gardiner, P.D., and Simmons, J.E. (1992). Analysis of conflict and change in construction projects, *Construction Management and Economics*, **10**(6): 459–478.

Ghalenoeei, N.K., Saghatforoush, E., Nikooravan, H.A., and Preece, C. (2021). Evaluating solutions to facilitate the presence of operation and maintenance contractors in the pre-occupancy phases: A case study of road infrastructure projects, *International Journal of Construction Management*, **21**(2): 140–152.

Jensen, P.A. (2009). Design integration of facilities management: A challenge of knowledge transfer, *Architectural Engineering and Design Management*, **5**(3): 124–135.

Kaplan, A.M., and Haenlein, M. (2010). Users of the world, unite! The challenges and opportunities of Social Media, *Business Horizons*, **53**(1): 59–68.

Karim, H., and Magnusson, R. (2008). Road design for future maintenance problems and possibilities, *Journal of Transportation Engineering*, **134**(12): 523–531.

Kozinets, R.V., Dolbec, P., and Earley, A. (2014). Netnographic analysis: Understanding culture through social media data, *The SAGE Handbook of Qualitative Data Analysis*, 262–276, U. Flick (Ed.), Sage, London.

Kumaraswamy, M.M., and Morris, D.A. (2002). Build-operate-transfer-type procurement in Asian megaprojects, *Journal of Construction Engineering and Management*, **128**(2): 93–102.

Lee, D., and Pan, R. (2017). Predictive maintenance of complex system with multi-level reliability structure, *International Journal of Production Research*, **55**(16): 4785–4801.

Lehtinen, J. (2021). *External Stakeholder Engagement in Complex Projects*, Aalto University.

Lin, H., Zeng, S., Ma, H., Zeng, R., and Tam, V.W.Y. (2017). An indicator system for evaluating megaproject social responsibility, *International Journal of Project Management*, **35**(7): 1415–1426.

Liu, J., Love, P.E., Smith, J., Regan, M., and Davis, P.R. (2014). Life cycle critical success factors for public-private partnership infrastructure projects, *Journal of Management in Engineering*, **31**(5): 401–407.

Liu, R., and Issa, R.R. (2016). Survey: Common knowledge in BIM for facility maintenance, *Journal of Performance of Constructed Facilities*, **30**(3): 1–8.

Liu, Y., van Marrewijk, A., Houwing, E.-J., and Hertogh, M. (2019). The co-creation of values-in-use at the front end of infrastructure development programmes, *International Journal of Project Management*, **37**(5): 684–695.

Lobo, S., and Abid, A.F. (2020). The role of social media in intrastakeholder strategies to influence decision making in a UK infrastructure megaproject: Crossrail 2, *Project Management Journal*, 8756972819864456.

Ma, G., Jia, J., Ding, J., Wu, M., and Wang, D. (2021). Examining the impact of social media use on project management performance: Evidence from construction projects in China, *Journal of Construction Engineering and Management*, **147**(3): 04021004.

Ma, H., Zeng, S., Lin, H., Chen, H., and Shi, J.J. (2017). The societal governance of megaproject social responsibility, *International Journal of Project Management*, **35**(7): 1365–1377.

Mathur, S., Ninan, J., Vuorinen, L., Ke, Y., and Sankaran, S. (2021). An exploratory study of the use of social media to assess benefits realisation in transport infrastructure projects, *Project Leadership and Society*, **2**: 1–10.

Mathur, S., Sankaran, S., MacAulay, S., and Tsang, I. (2021). A framework to manage data science initiatives, *The 6th PMI Research and Academic Virtual Conference 2021*, 30–49.

Meng, X., Zhao, Q., and Shen, Q. (2011). Critical success factors for transfer-operate-transfer urban water supply projects in China, *Journal of Management in Engineering*, **27**(4): 243–251.

Meng, X. (2013). Involvement of facilities management specialists in building design: United Kingdom experience, *Journal of Performance of Constructed Facilities*, **27**(5): 500–507.

Mergel, I., Rethemeyer, R.K., and Isett, K. (2016). Big data in public affairs, *Public Administration Review*, **76**(6): 928–937.

Mladenovic, G., Vajdic, N., Wündsch, B., and Temeljotov-Salaj, A. (2013). Use of key performance indicators for PPP transport projects to meet stakeholders' performance objectives, *Built Environment Project and Asset Management*, **3**(2): 228–249.

Mok, K.Y., Shen, G.Q., and Yang, J. (2015). Stakeholder management studies in mega construction projects: A review and future directions, *International Journal of Project Management*, **33**(2): 446–457.

Ninan, J. (2020). Online naturalistic inquiry in project management research: Directions for research, *Project Leadership and Society*, **1**: 100002.

Ninan, J., Clegg, S., and Mahalingam, A. (2019). Branding and governmentality for infrastructure megaprojects: The role of social media, *International Journal of Project Management*, **37**(1): 59–72.

Ninan, J., Mahalingam, A., and Clegg, S. (2020). Power and strategies in the external stakeholder management of megaprojects: A circuitry framework, *Engineering Project Organization Journal*, **9**(1): 1–20.

Ninan, J., Mahalingam, A., and Clegg, S. (2021). Asset creation team rationalities and strategic discourses: evidence from India, *Infrastructure Asset Management*, **8**(2): 1–10.

Oldenhof, L., Postma, J., and Putters, K. (2014). On justification work: How compromising enables public managers to deal with conflicting values, *Public Administration Review*, **74**(1): 52–63.

Osei-Kyei, R., and Chan, A.P.C. (2015). Review of studies on the critical success factors for public-private partnership (PPP) projects from 1990 to 2013, *International Journal of Project Management*, **33**(6): 1335–1346.

Osei-Kyei, R., Chan, A.P., and Ameyaw, E.E. (2017). A fuzzy synthetic evaluation analysis of operational management critical success factors for public-private partnership infrastructure projects, *Benchmarking: An International Journal*, **24**(7): 2092–2112.

Park, C.S. (2015). Applying 'Negativity Bias' to twitter: Negative news on twitter, emotions, and political learning, *Journal of Information Technology & Politics*, **12**(4): 342–359.

Partnerships UK. (2006). *PFI Strengthening Long Term Partnerships*, HM Treasury, London.

Provost, F., and Fawcett, T. (2013). Data science and its relationship to big data and data-driven decision making, *Big Data*, **1**(1): 51–59.

Raghupathi, W., and Raghupathi, V. (2014). Big data analytics in healthcare: Promise and potential, *Health Information Science and Systems*, **2**(1): 1–10.

Rivera-Vazquez, J.C., Ortiz-Fournier, L.V., and Flores, F.R. (2009). Overcoming cultural barriers for innovation and knowledge sharing, *Journal of Knowledge Management*, **13**(5): 257–270.

Robinson, H.S., and Scott, J. (2009). Service delivery and performance monitoring in PFI/PPP projects, *Construction Management and Economics*, **27**(2): 181–197.

Saunders, M., Lewis, P., and Thornhill, A. (2009). *Research Methods for Business Students*, 5th ed., Pearson Education Limited, Harlow.

Scott, W.R. (1965). Field methods in the study of organisations, *Handbook of Organizations*, 261–304, J.G. March (Ed.). Rand McNally and Co., Chicago.

Steenhuisen, B., and van Eeten, M. (2008). Invisible trade-offs of public values: Inside dutch railways, *Public Money & Management*, **28**(3): 147–152.

Steger, T. (2007). The stories metaphors tell: Metaphors as a tool to decipher tacit aspects in narratives, *Field Methods*, **19**(1): 3–23.

Tang, L., Shen, Q., Skitmore, M., and Cheng, E.W. (2012). Ranked critical factors in PPP briefings, *Journal of Management in Engineering*, **29**(2): 164–171.

van der Aalst, W. (2016). Data science in action. pp. 3–23. *In:* van der Aalst, W. (Ed.). *Process Mining*, Springer, Berlin.

Wang, G., Wu, P., Wu, X., Zhang, H., Guo, Q., and Cai, Y. (2020). Mapping global research on sustainability of megaproject management: A scientometric review, *Journal of Cleaner Production*, **259**: 120831.

Wang, S., and Xie, J. (2002). Integrating building management system and facilities management on the internet, *Automation in Construction*, **11**(6): 707–715.

Yin, R.K. (2017). *Case Study Research and Applications: Design and Methods*, Sage Publications, Thousand Oaks, California.

Ziora, A.C.L. (2015). The role of big data solutions in the management of organisations, review of selected practical examples, *Procedia Computer Science*, **65**: 1006–1012.

8

Mining Online Public Opinions on Megaprojects

A Case Study of The Hong Kong-Zhuhai-Macao Bridge

Zhipeng Zhou, Xingnan Zhou, Lingfei Qian* and *Haonan Qi*

8.1 Introduction

Compared to general infrastructures, large-scale infrastructures, usually called as mega-infrastructures, are characterized by huge amounts of investment, high degree of technological complexity, wide range of participators, extremely long life cycle and so on (Jia et al., 2011). Megaprojects are developed to carry out grand strategies of organizations, or to ambitiously make changes of societal structure (Flyvbjerg, 2014). Megaprojects, such as railway, airport, canal, bridge, water-supply system are closely concerned with individual livelihoods, and the public are exerting much more influence on the entire life cycle of a megaproject (Williams et al., 2015) from its beginning of plan, design, construction, operation, and maintenance to ex-post project evaluation. Hence, public opinions deserve enough attention in order to achieve high performance and final success of megaproject management.

The HZMB is the world's longest sea-crossing bridge and is located at the Pearl River Estuary in China. Opened on 24 October 2018, it cuts

Nanjing University of Aeronautics and Astronautics, China.
* Corresponding author: Zhouzhipeng@nuaa.edu.cn

the road travel time between Hong Kong and Zhuhai/Macau from three hours to merely 40–50 minutes. HZMB has the ability to enhance the connection of the bay area and boost exchange of personnel, goods, information, and capital between the three places. Not only for enhancing the economic and sustainable development of the three places, this mega-size sea-crossing bridge will be a magnet to attract more resources in the Greater Bay Area and become a stronger growth driver for the region, even for the whole southern part of China. Hence, HZMB gains extensive focus not only from the domain of bridge design and construction, but also from public community. Although the general public may not be familiar with the detailed process of HZMB's development, the final deliverable is directly related to their life, work or business, especially for local people or organizations.

This study tries to develop an online public opinion evolution analysis framework, which combines topic extraction and sentiment analysis through crawling online data from SINA Weibo. In the aspect of topic extraction, it determines the optimal number of topics by the index of perplexity at each stage and extracts corresponding topics of microblog data based on latent Dirichlet allocation (LDA). In the aspect of sentiment analysis, it integrates existing dictionaries and high frequency words pertinent to HZMB public opinion management to establish an extensive sentiment dictionary. It is used to identify the polarity of the comments and calculate the values of sentiments. According to milestone events in the development of HZMB, the evolution process of the two dimensions of topics and sentiments is illustrated across various stages of this megaproject.

8.2 Literature Review

Classics are essential for an academic field to make progress (Kuhn, 2012), and megaproject management is no exception. Professor Bent Flyvbjerg has a high reputation in the field of megaproject management, working as an advisor on questions of megaprojects and risk. In order to balance his possible idiosyncrasies, Flyvbjerg and Turner (2018) carried out a survey and invited 114 experts to list most influential and classical articles in the area of megaproject management. A bibliographic meta-network analysis for the most cited classic literature in five selected management theories as a reference group was conducted to provide a potential path for the evolution of megaproject management theory (Li et al., 2018). Siemiatycki (2018) took one article as an example to identify key ingredients which make a classic text. On the basis of three major books (Flyvbjerg et al., 2003; Morris and Hough, 1988; Merrow, 2011), Pollack (2018) deployed four criteria for a structured analysis determining whether a book can be regarded to be classical.

Online public opinion management often involves two main aspects of topic analysis and sentiment analysis in the area of natural language processing technology. Topic analysis is a machine learning technique to identify the frequent topics from large collections of unstructured data (O'Callaghan et al., 2015). Topics which are deemed as patterns across datasets are crucial to the description of an issue or a phenomenon. Topic classification and modelling can be utilized to make sense of seemingly uncorrelated data (Zirn and Stuckenschmidt, 2019). Among the algorithms for topic modelling, the most popular is latent Dirichlet allocation (LDA), which is a three-layer hierarchical Bayesian model of word, topic, and document (Blei et al., 2003). LDA model is widely used for its clear structure, high efficiency, and accuracy (Chen, 2017). LDA is an unsupervised machine learning technique which can be utilized to identify hidden topic information in large-scale document collections or corpus.

Sentiment analysis is a way to evaluate written or spoken language to determine whether the expression is favorable, unfavorable, or neutral (Maks and Vossen, 2012). There are two basic approaches for sentiment analysis. One approach is rule-based, referring to the research implemented by the language experts (Maks and Vossen, 2012). In the beginning of sentiment analysis, a set of rules, known as sentiment lexicon or dictionary, should be constructed (Wu et al., 2016). Another approach is based on machine learning which is more automatic than the rule-based approach (Ghiassi and Lee, 2018). Instead of clearly defining rules, this approach utilizes supervised or unsupervised machine learning classification algorithms to calculate the polarity and intensity of words (Fernández-Gavilanes et al., 2016). The following algorithms of naïve Bayes, support vector machine, long short-term memory, and random forest are usually employed for emotional data exploration.

In case of the proliferation of social media platforms, people realize that public opinion analysis is changed a lot instead of depending on traditional survey research (Lee and Chun, 2016). A variety of social media platforms are often touted as important influencers on public opinion affecting various fields (Kirilenko and Stepchenkova, 2014). Definitely there is no exception for project management, especially for megaproject management which has a higher impact on social development and the interests of public community. Compared to conventional project research areas, such as time estimation, cost control, safety risk, change management, and performance evaluation, there are fewer studies focusing on public opinion analysis and management.

The infrastructure of Three Gorges Dam (TGD) is the world's biggest hydropower which contributes to society in different aspects, such as water reservation, flood prevention, and power generation (Shen, 2004). However, it has also resulted in environmental disruption, landscape loss, migration issue, geological hazard, governmental corruption (Tullos, 2009). General public maintains concentration on the controversial mega-infrastructure. A project sentiment analysis model based on lexicon approach was proposed to obtain public opinions from SINA Weibo, establish emotional dictionary, and determine basic rules for emotion computation (Jiang et al., 2016). Instead of public opinion analysis on individual project, Tang et al. (2017) considered construction as a project-based industry. An integrated framework of sentiment analysis, topic modelling, link analysis, and geographical location analysis based on data from Twitter was developed to find the interactions among the four types of users in construction. It indicated that construction labors tended to have a higher proportion of negative messages compared to the other clusters. One case study was conducted to examine the network structure of online stakeholder discussions in the early stage of a public megaproject of High-Speed Rail Two in the United Kingdom (Williams et al., 2015).

8.3 Methodology

Opinions from the public will vary during a megaproject's long life cycle. According to construction milestones of HZMB, its total life cycle can be separated into four stages. The first stage is from the beginning to 7 July 2017, when the main structure of HZMB was completed. The second stage is from 8 July 2017 to 28 December 2017, when the deck pavement of HZMB was completed. The third stage is from 29 December 2017 to 23 October 2018, when the ceremony was held to mark the full completion and prepare for official operation of HZMB. The fourth stage is the operation stage from 24 October 2018, to present. Hence this research aims to explore varied public opinions along these four stages of HZMB.

8.3.1 Data Collection

Social media platforms provide user-generated content including posts, comments, and so on through online interactions. It is beneficial to capture a large quantity of public opinions on a particular subject via social media networks. Therefore, this study adopts social media platform for collecting raw data of public opinions which are limited in the scope of China. There are many social media platforms, such as WeChat and SINA Weibo in Chinese with a lot of netizens. Given that SINA Weibo is a highly popular microblogging social platform known for its hybrid mix of

Twitter's and Facebook's features, we utilized SINA Weibo as the platform for gathering data.

On the basis of the Application Programming Interface (API), a web crawler was developed to browse at SINA Weibo website to obtain raw data pertinent to the HZMB. The keyword is Hong Kong-Zhuhai-Macao Bridge. In the four stages, netizens posted 743, 776, 978, 5515 microblogs and 11462, 4845, 3621, 23809 comments respectively. Consequently, a total of 51,749 raw data on the HZMB megaproject were obtained. Among them there are 8,012 microblogs and 43,737 comments. Besides microblogs and comments, the information related to blogger ID, posting date and time was also involved in final dataset.

8.3.2 Data Pre-processing

Due to incomplete and inconsistent data from the crawling approach, data pre-processing is a requisite previous to data analysis. Firstly, invalid data were eliminated. The data with blank value, duplicating value, and litter value should be removed. Additionally, the data which merely have the word of Forward Weibo or which are relative to HZMB's advertisement, are considered to be invalid. Secondly, noisy texts were eliminated. The useless information of nickname often appears after the symbol of @ in blog posts and comments. If a nickname appears frequently, it may affect the topic extraction results, and the emotional information contained in the nickname may also have an impact on the sentiment-analysis results of comments. Some punctuations, such as '#Text#' and '[Text]' which are adopted to mark a topic should be deleted. Thirdly, picture information was eliminated. This study merely analyzes the topic and sentiment on the basis of plain texts. Hence the image information and emoji information from the blog posts and comments should be removed.

As a result, a total of 41,846 pre-processed data, inclusive of 7,003 blogs and 34,843 comments, were validated based on the social media network of SINA Weibo.

8.3.3 Data Analysis

General approach to determine whether two documents are similar is to calculate the number of words which appear together in these two documents, such as TF-IDF. This approach usually does not take into account the semantic association or meaning underlying the words in a document. For example, the Sentence 1: 'Steve Jobs died' and the Sentence 2: 'Will the price of iPhone go down due to the change of CEO?' do not have common words, but the two sentences have an internal link. If the two sentences are judged in a traditional way, they must not be similar. Document semantics cannot be neglected in the process of judging similarity and relevance of documents.

The LDA model considers each word of an article to be obtained through a process of selecting a topic with a certain probability and selecting a word from this topic with a certain probability (Puranam et al., 2017). When the text words are already known, the topic distribution of the text set can be deduced through a probabilistic method (Wang and Xu, 2018). Accordingly, the document can be clustered. It is suitable for the modelling of massive heterogeneous text data, which greatly reduces the dimensionality of text representation (Blei et al., 2003). K is the number of topics, M is the number of documents, and N is the number of words. For the nth word of the mth document, the topic distribution corresponding to the document is first generated by the prior distribution α (Dirichlet distribution) of the topic, and a topic Zn is extracted therefrom. According to the current topic Zn, the distribution ϕk of the word is generated from the prior distribution β (Dirichlet distribution) of the word, and a word Wn is extracted from the distribution. After N times of repeats of previous process, it will get a document with N words. Formula (1) illustrates the joint probability of LDA.

$$p(\theta, z, w \mid \alpha, \beta) = p(\theta \mid \alpha) \prod_{n=1}^{N} p(z_n \mid \theta) p(w_n \mid z_n, \beta) \tag{1}$$

The indicator of perplexity will be adopted to determine the optimal number of topics in LDA topic modelling. Perplexity is a measurement of how well a probability distribution or probability model predicts a sample. It can also be used to compare two probability distributions or probability models. The probability distribution model or probability model with low degree of confusion can better predict the sample. Perplexity can be divided into three categories of a probability distribution, a probability model, and per word. In natural language processing technology, the last one is usually selected to measure the effectiveness of language probability model (Toral et al., 2015). Accordingly, the value of perplexity in LDA model is calculated as below.

$$p(\vec{w}_{\mathrm{m}} \mid M) = \prod_{n=1}^{N_m} \prod_{k=1}^{K} p(w_n = t \mid z_{n=k}) \cdot p(z_n = k \mid d = m) = \prod_{t=1}^{V} \left(\sum_{k=1}^{K} \phi_{k,t} \cdot \theta_{m,k} \right)^{n_m^{(t)}} \tag{2}$$

We combined the sentiment dictionary from Information Retrieval Laboratory of Dalian University of Technology (DUTIR) with BosonNLP which focuses on network words, in order to extract high-frequency words from the comments relative to the megaproject of HZMB. If a high-frequency word is not found in the original sentiment dictionary, it will be added to update the dictionary and corresponding value will be assigned to the word, according to its synonyms. For example, the term *infrastructure maniac* appears many times in the comments on the megaproject, but not in the original dictionary. This high-frequency word means China is excellent in infrastructure construction and development. Accordingly, the value of *infrastructure maniac* is similar to the value of excellent. In addition, there are some words with opposite meanings in the original dictionary. For example, hard work is a negative sentimental word in the original dictionary. In this situation, it represents praise for the spirit of HZMB construction workers, so its sentimental value is reversed.

The sentiment value of a comment is often influenced by three types of words of emotion, degree, and negative words. Firstly, it is to extract emotional words, degree words, and negative words from a comment, and identify their corresponding positions in the comment. Secondly, it is to calculate the value of an emotional word on the basis of relative positions among the emotional word, degree word, and negative word. Calculation rules proposed by Cao et al. (2014) are adopted for value determination. When the relative position is degree word + sentiment word or sentiment word + degree word, calculation formula is:

$$V = SV \times DV \tag{3}$$

where V is the value of the emotional word, SV is the intensity of the emotional word, and DV is the intensity of corresponding degree word. When the relative position is negative word + sentiment word, calculation formula is as follows:

$$V = -SV \tag{4}$$

Formula (5) illustrates the value when the relative position is negative word + degree word + sentiment word.

$$V = -1/3 \times SV \times DV \tag{5}$$

The last case is degree word + negative word + sentiment word, the sentiment value is calculated as below:

$$V = -SV \times DV \tag{6}$$

Thirdly, activities in step 2 are repeated to calculate the values of other emotional words in the comment. Fourthly, the value of the comment (V_{total}) is obtained by adding up the values of all emotional words. If V_{total} is greater than 0, the comment is regarded to be positive. If V_{total} is equal to 0, the comment is regarded to be neutral. If V_{total} is smaller than 0, the comment is regarded to be negative.

8.4 Results and Discussion

8.4.1 Topics Extraction

Given that Stage 1 is mostly related to the construction of HZMB, public community had paid much attention to the milestones and issues in the process of bridge construction. The immersed tunnel beneath the Pearl River Estuary was deemed to be the most challenging section of HZMB as the longest sea-crossing bridge in the world. People were focusing on quantity of tunnel elements, heavy wedge installation, joint welding and so on. Eastern and western artificial islands were constructed to offer a smooth transition between the section above the water and the undersea tunnel.

Despite the fact that the public were supporting and expecting this megaproject, they also showed a variety of concerns for schedule delay, safety accident, engineering fraud, etc. The necessary environmental assessment report did not contain the impact analysis of ozone, carbon dioxide, sulphur dioxide and suspended particles resulting from the construction of HZMB. Hong Kong High Court declared it illegal and invalid, which led to the requirement for a new environmental assessment report. HZMB faced multiple challenges, such as harsh construction environment which increased the safety risks inevitably. As a result, various accidents inclusive of fall from height, collapse, struck by falling objects, caught in or between, and drowning had happened on site. Corruption or fraud has become a major issue in the construction project management, especially for this mega-bridge with large investment and technological complexity.

In Stage 2, people were surprised by the high productivity of deck pavement work. A total of 7,000 square meters deck pavement of HZMB in the section of Zhuhai merely took eight days for completion. The entire pavement of this mega-bridge covered an area of 700,000 square meters equivalent to 98 football fields. It involved paving water-resistant and porous asphalt layers to shield the steel structure of the bridge from high levels of relative humidity in the Great Bay Area. Artificial island project is an integral part of the large infrastructure because there are limited lands of artificial islands, complex functional requirements, and people-oriented design for affiliated facilities are necessary. As HZMB's construction was close to the end, the public were more confident about its future function and success in Great Bay Area. Simultaneously, people are proud of this new megaproject with technological complexity.

In Stage 3, it was seen that the new bridge with a total length of 55 kilometers was illuminated on the last day of 2017, in order to mark the successful installation of the lighting system with reliable power supply. There were stunning fireworks together with the lighting-up ceremony. Labours and experts had conquered the difficulties of complex construction environment and the lack of an optical fibre to provide tailored 4G solutions for the bridge and prepare for the future upgrading of 5G service. Public community also focused on precise date for officially starting service of HZMB. Partially relying on the development progress of port facilities in Zhuhai and Hong Kong, the final commissioning date was scheduled on 24 October 2018.

As the mega-bridge was about to open to traffic, detail policies for cross-border traffic among three local governments of Hong Kong, Zhuhai and Macau were discussed by the public. There are visa issues for the residents from both China and other nations or regions. Because each region is self-governed, each has its own vehicle registration and licensing process. The company of One-Bus would offer 24-hour shuttle service between Hong Kong and Macau through HZMB. Peak period headway is around 5 minutes, non-peak period headway is from 10 to 15 minutes, and overnight period headway is from 15 to 30 minutes. Cross-boundary car owners are required for obtaining closed road permit issued by the transport department of Hong Kong, identification label issued by the government of Macau, and electronic temporary vehicle license from the Mainland.

In Stage 4, during the Spring Festival of 2019, many tourists paid a visit to HZMB which has become an instagrammable place. Online users sent a variety of congratulatory messages from all parts of the world. Also, tens of thousands of workers and managers have interpreted the spirit of hard work. This megaproject is considered as one of the seven new miracles of the world, according to the report from *The Guardian*.

There are multiple functions of the bridge. It is part of a planned national highway network linking the western and eastern banks of the Pearl river. HZMB cuts what was a three-hour trip between Zhuhai and Hong Kong down to about 40–50 minutes and gives both residents and travellers a quicker and more efficient way to move among Hong Kong, the mainland and Macau. The new landmark of HZMB underscores Hong Kong as a vibrant gateway for the western part of Pearl river estuary. It is also the closest gateway for cross-border visitors to the Hong Kong Disneyland Park from both the mainland and Macau. The uniqueness of HZMB brings the three cities of Hong Kong, Zhuhai and Macau much closer. Not only as a linkage of the three cities, this mega infrastructure links '9 + 2' cities of the Greater Bay Area into a giant megalopolis with high population density. The bridge injects new vigor into the area's logistics and tourism. The Ministry of Transport in China released that the total passenger capacity was 1.79 million after the operation of HZMB for the first month. According to statistics, the total number of visitors who enter the HZMB daily stands at 64,000. It is estimated that the potential daily contribution of the bridge to the Macau economy is Macau Pataca 16.4 million. Since the opening of HZMB for almost one year, the Greater Bay Area has become more integrated and shows stronger synergistic effect.

With the operation of HZMB, the premiere of the documentary HZMB was held in Zhuhai. It has two episodes, each lasting 50 minutes. The documentary crew travelled to different countries, such as the United States of America, Bangladesh and Brunei to explore the bridge's value and influence against the backdrop of the Belt and Road Initiative. While HZMB has opened for more than one year, there are limited vehicles driving on the bridge. The utilization rate of HZMB is not high. Due to different rules for transportation among Hong Kong, Zhuhai and Macau, not all the vehicles can drive through this mega-bridge which is still in its beginning of operation. HZMB's political significance is more obvious than its economic significance at present. The ongoing protests in Hong Kong have also contributed to a loss of 1.9 billion Yuan in HZMB's first year of operation, with traffic less than half the official predictions.

8.4.2 Sentiment Analysis

The sentiment analysis of pre-processed comment dataset was implemented on the basis of the extended sentiment dictionary for the megaproject of HZMB. In order to assess effectiveness of the extended sentiment dictionary, partial comment data were randomly chosen for manual tagging. Considering that the number of comment data for each stage is discrepant, 772, 734, 186, and 1,788 comment data were selected from the four stages, respectively, according to the proportion of comment data at each stage. As a result, a total of 3,480 comment data accounting for 10% of all were obtained as a testing dataset.

Classification of the testing dataset produces a confusion matrix with four outcomes of true positive (TP), false positive (FP), true negative (TN), and false negative (FN) (Powers, 2011). True positives and true negatives are the observations which were correctly predicted. False positives are observations where the actual was negative. False negatives are observations where the actual was positive. Various metrics can be derived from the confusion matrix. Six metrics of accuracy—precision (P), precision (N), recall (P), recall (N), and F1-score—were adopted for the evaluation on the extended sentiment dictionary. Accuracy is perhaps the most intuitive performance metric. It is simply the ratio of correct identification of all the testing dataset. Precision (P) is the ratio of correct positive identifications to the total identified positives. Precision (N) is the ratio of correct negative identifications to the total identified negatives. Recall (P) is the ratio of correct positive identifications to the total positives of all the testing dataset. Recall (N) is the ratio of correct negative identifications to the total negatives of all the testing dataset. F1-score, which takes both false positives and false negatives into account, is the weighted average of the metrics of precision and recall. The formulas (7)–(12) show how to calculate these six-assessment metrics as below.

$$Accuracy = \frac{TP + TN}{TP + FP + FN + TN} \tag{7}$$

$$\text{Precision } (P) = \frac{TP}{TP + FP} \tag{8}$$

$$\text{Recall } (P) = \frac{TP}{TP + FN} \tag{9}$$

$$\text{Precision } (N) = \frac{TN}{TN + FN} \tag{10}$$

$$\text{Recall } (P) = \frac{TN}{TN + FP} \tag{11}$$

$$F1 - score = \frac{2 \times \text{Precision} \times \text{Recall}}{\text{Precision} \times \text{Recall}} \tag{12}$$

From the calculation results of assessment metrics of 3,480 comment data, the testing dataset based on the extended sentiment dictionary has the high accuracy rate of 94.1%. It means that only 5.9% of sentiment results are not correct. The values of precision (P) and precision (N) are 96.2% and 85.6%, respectively. Precision (P) is higher than accuracy, but precision (N) is lower than accuracy. That indicates false positive sentiments are a smaller part of the error set, compared to false negative sentiments. Recall is the opposite of precision. The values of recall (P) and recall (N) are 96.4% and 84.6%, respectively. Comparing to the values of precision (P) and precision (N), precision and recall are pretty similar in both positive and negative sentiments. There is a trade-off between precision and recall. F1-score is the harmonic mean of precision and recall to accomplish the trade-off between precision and recall. The proposed approach has an F1-score of 96.3% in positive perspectives and 85.1% in negative perspectives, but the value of F1-score in positive perspective is around 10% higher than the one in negative perspective. This means the extended sentiment dictionary recognizes positive sentiments better. On the whole, the values of these assessment metrics reveal that the proposed approach is better for sentiment identification and meets the requirements for sentiment analysis.

Table 8.1: Proportion of Different Types of Comments in Four Stages.

Stage	Stage 1	Stage 2	Stage 3	Stage 4
Negative comments	0.27	0.19	0.22	0.18
Neutral comments	0.05	0.04	0.05	0.08
Positive comments	0.68	0.77	0.73	0.74

The sentiment analysis of the pre-processed comment dataset was conducted on the basis of calculation rules in Chapter 3. According to sentiment values, comments are divided into three types—positive, neutral and negative. If the value is greater than zero, it means corresponding comment expresses positive sentiment. If the value is lower than zero, it means corresponding comment expresses negative sentiment. If the value is equal to zero, it means corresponding comment is neutral. Table 8.1 exhibits the proportion of positive, neutral, negative, and non-negative comments across four stages of HZMB.

The values of comment data in every type were added together to obtain intensity values of every type's comments. Sentiment intensity of a comment indicates the strength of its association with the sentiment. For the positive comments of this mega-infrastructure, the total intensity value of the four stages is 24304, 11237, 8607, 38714 respectively. In order to remove the influence of the number of comments, this study calculates the mean value, which are 3.44, 3.47, 3.37, 3.12. In the same way, the total intensity value of each stage of negative comments are –6389, –1811, –1398, –4933, and the average value are –2.24, –2.23, –1.86, –1.63.

8.4.3 Discussion

In contrast to the other three stages, there was the highest percentage of negative comments accounting for 27% of all in the first stage of HZMB. Combined with the results from topic extraction, it is found that various dissatisfying events happened in the stage, such as construction safety accidents, falsification of engineering testing, illegal environment assessment report, and schedule delay. There were 68% positive sentiments which were mainly from the official dissemination of HZMB's functions and significance for social and economic development.

With the connection of principal project of HZMB, all media outlets showered praises and the new mega-infrastructure got recognition all over the world in multiple aspects, such as longest sea-crossing bridge, undersea tunnel with the length of 6.7 kilometers, deepest immersed tube beneath the water, and technical and managerial innovation. These commendations made the proportion of positive comments rise to 77%, and the proportion of negative comments fall by 8%.

In the third stage, due to the arrival of typhoon Mangkhut, public community felt worried whether HZMB could withstand the super typhoon with a maximum wind speed of 55 m/s. In addition, drone photos of one artificial island resulted in public doubts on the quality of breakwaters to some degree. Therefore, there was a slight increase of negative sentiments by 3%. Compared to the first two stages, the average intensity value of negative comments decreased from around –2.23 to –1.86 in this stage. This indicated that there were a smaller number of negative issues which might lead to intense negative sentiments. People were inclined to worry about the reliability of this megaproject, since it was very close to opening to traffic.

After official operation, people expressed their praise for shortening the driving time among the three cities of Hong Kong, Zhuhai and Macau. The 70th anniversary of founding of the People's Republic of China also stimulated the patriotism from netizens, and the proportion of positive comments increased from 73% in the third stage to 74% in the stage of operation. However, there were still 17% negative comments of all in the stage and these were mainly manifested in dissatisfaction with the inconvenience of license plate and low traffic flow through the large bridge.

As to sentiment intensity, the larger absolute value of average sentiment intensity means a higher certainty and a clearer polarity of sentiments. Average intensity value of positive comments was larger than the absolute value of average intensity of negative comments across four stages of HZMB. It denoted that positive comments were disseminated in a relatively smooth path, and negative comments were well controlled in an effective and efficient way. In spite of some negative incidents which occurred in the process of developing this megaproject, the public still keep a high satisfaction and a strong expectation of HZMB's success in its future.

On the basis of extensive analyses and discussion, some managerial implications are offered as below. Firstly, public opinions of a mega-infrastructure need monitoring in time, especially when accidents or deviations occur. In the process of HZMB, public negative sentiments surge sharply due to safety accidents and fraud issues. In case of these negative events, the management team should make better use of the influence of social media to release the detail information of the events in time, so that the public can access and understand the real causes and consequences of the events. It helps avoid the outbreak of public dissatisfaction caused by the spread of possible rumors. Secondly, extensive attention should be paid to the influence of public opinions in early stages on later stages. As can be seen from the case, there was certain continuity in the topic of public opinions at each stage. Public concerns in planning or construction stages may be raised again in the later operation and maintenance stage.

Public opinion analysis in the early stage contributes to predicting the topic and sentiment trend in the later stage, so as to guide public opinion management in a proactive way. Thirdly, although the approach for online public opinion was proposed on the basis of the case of HZMB, it can also be applied to other infrastructure projects, such as highway, high-speed railway, and nuclear power. It is possible to be extended to multiple cases to explore the similarities of public concerns and sentimental changes. This will be beneficial for summarizing the characteristics of public opinion changes of various projects, so that project managers or government departments have the potential to quickly respond to public sentiments when encountering a similar situation.

8.5 Conclusion

This study proposed an online public opinion evolution analysis framework which collected raw data from SINA Weibo. HZMB as the world's longest sea-crossing bridge was selected as the case for detailed demonstration of the framework. Based on web crawler technology, the data were obtained across various stages of this mega-infrastructure. A three-step approach was utilized for data pre-processing. Consequently, a total of 41,846 pre-processed data inclusive of 7,003 microblogs and 34,843 comments remained. As a flexible generative probabilistic model, latent Dirichlet allocation (LDA) was adopted for topic extraction which was carried out separately in HZMB's four phases. The index of perplexity was used to determine the optimal number of topics in each stage. Furthermore, lexicon-based sentiment analysis was conducted for the calculation of sentiment values of 34,843 comment data. The evolution of public sentiment was displayed from the total life cycle of HZMB.

Corresponding results show that public concerns and their changes are across different stages of HZMB. Average intensity value of positive sentiment was larger than the absolute value of average intensity of negative sentiment for each stage of HZMB. Online public opinion analysis is helpful for the perception of a variety of topics on the new bridge. Managers and their team members are possible to implement topic prediction and sentiment warning for HZMB's following phase. Managerial implications can be recommended as monitoring and responding to public opinions in time, early concerns on public opinions at the beginning of total life cycle, and extensive application of public opinion analysis approach in similar megaprojects.

References

Blei, D.M., Ng, A.Y., and Jordan, M.I. (2003). Latent Dirichlet allocation, *Journal of Machine Learning Research*, **3**: 993–1022.

Cao, J., Zeng, K., Wang, H., Cheng, J., Qiao, F., Wen, D., and Gao, Y. (2014). Web-based traffic sentiment analysis: Methods and applications, *IEEE Transactions on Intelligent Transportation System*, **15**(2): 844–853.

Chen, L. (2017). An effective LDA-based time topic model to improve blog search performance, *Information Processing & Management*, **53**(6): 1299–1319.

Fernández-Gavilanes, M., Álvarez-López, T., Juncal-Martínez, J., Costa-Montenegro, E., and González-Castaño, J.F. (2016). Unsupervised method for sentiment analysis in online texts, *Expert Systems with Applications*, **58**: 57–75.

Flyvbjerg, B., Bruzelius, N., and Rothengatter, W. (2003). *Megaprojects and Risk: An Anatomy of Ambition*, Cambridge University Press, Cambridge, U.K.

Flyvbjerg, B. (2014). What you should know about megaprojects and why: An overview, *Project Management Journal*, **45**(2): 6–19.

Flyvbjerg, B., and Turner, J.R. (2018). Do classics exist in megaproject management, *International Journal of Project Management*, **36**(2): 334–341.

Ghiassi, M., and Lee, S. (2018). A domain transferable lexicon set for twitter sentiment analysis using a supervised machine learning approach, *Expert Systems with Applications*, **106**: 197–216.

Jia, G., Yang, F., Wang, G., Hong, B., and You, R. (2011). A study of mega project from a perspective of social conflict theory, *International Journal Project Management*, **29**(7): 817–827.

Jiang, H., Lin, P., and Qiang, M. (2016). Public opinion sentiment analysis for large hydro projects, *Journal of Construction Engineering and Management*, **142**(2): 05015013-1-12.

Kirilenko, A.P., and Stepchenkova, S.O. (2014). Public microblogging on climate change: One year of twitter worldwide, *Global Environmental Change*, **26**: 171–182.

Kuhn, T.S. (2012). *The Structure of Scientific Revolutions* (4th ed.), University of Chicago Press, Chicago, USA.

Lee, M.J., and Chun, J.W. (2016). Reading others' comments and public opinion poll results on social media: Social judgment and spiral of empowerment, *Computers in Human Behavior*, **65**: 479–487.

Li, Y., Lu, Y., Taylor, J. E., and Han, Y. (2018). Bibliographic and comparative analyses to explore emerging classic texts in megaproject management, *International Journal of Project Management*, **36**(2): 342–361.

Maks, I., and Vossen, P. (2012). A lexicon model for deep sentiment analysis and opinion mining applications, *Decision Support Systems*, **53**(4): 680–688.

Merrow, E.W. (2011). *Industrial Megaprojects: Concepts, Strategies and Practices for Success* (1st ed.), John Wiley, New Jersey, USA.

Morris, P., and Hough, G.H. (1988). *The Anatomy of Major Projects: A Study of the Reality of Project Management* (1st ed.), John Wiley, New Jersey, U.S.A.

O'Callaghan, D., Greene, D., Carthy, J., and Cunningham, P. (2015). An analysis of the coherence of descriptors in topic modelling, *Expert Systems with Applications*, **42**(13): 5645–5657.

Pollack, J., Biesenthal, C., Sankaran, S., and Clegg, S. (2018). Classics in megaproject management: A structured analysis of three major works, *International Journal of Project Management*, **36**(2): 372–384.

Powers, D.M.W. (2011). Evaluation: From precision, recall and f-measure to ROC, informedness, markedness and correlation, *Journal of Machine Learning Technologies*, **2**(1): 37–63.

Puranam, D., Narayan, V., and Kadiyali, V. (2017). The effect of calorie posting regulation on consumer opinion: A flexible latent Dirichlet allocation model with informative priors. Marketing Science, **36**(5): 726–746.

Shen, G. (2004). Three gorges project: Chance and challenge, *Science*, **304**(5671): 681–681.

Siemiatycki, M. (2018). The making and impacts of a classic text in megaproject management: The case of cost overrun research, *International Journal of Project Management*, **36**(2): 362–371.

Tang, L., Zhang, Y., Dai, F., Yoon, Y., Song, Y., and Sharma, R.S. (2017). Social media data analytics for the US construction industry: Preliminary study on Twitter, *Journal of Management in Engineering*, **33**(6): 04017038-1-15.

Toral, A., Pecina, P., Wang, L., and van Genabith, J. (2015). Linguistically-augmented perplexity based data selection for language models, *Computer Speech & Language*, **32**(1): 11–26.

Tullos, D. (2009). Assessing the influence of environmental impact assessments on science and policy: An analysis of the three gorges project, *Journal of Environmental Management*, **90**(S3): s208-s223.

Wang, Y., and Xu, W. (2018). Leveraging deep learning with LDA-based text analytics to detect automobile insurance fraud, *Decision Support Systems*, **105**: 87–95.

Williams, N.L., Ferdinand, N., and Pasian, B. (2015). Online stakeholder interactions in the early stage of a megaproject, *Project Management Journal*, **46**(6): 92–110.

Wu, F., Huang, Y., Song, Y., and Liu, S. (2016). Towards building a high-quality microblog-specific Chinese sentiment lexicon, *Decision Support Systems*, **87**: 39–49.

Zirn, C., and Stuckenschmidt, H. (2014). Multidimensional topic analysis in political texts, *Data & Knowledge Engineering*, **90**: 38–53.

9

Project Organizations' Roles in Using Social Media for External Stakeholder Engagement

Implications on Value Creation and Distribution

Jere Lehtinen[1,*] and *Kirsi Aaltonen*[2]

9.1 Introduction

Projects, whose aim is always some kind of change, affect a range of external stakeholders (Mok et al., 2015). We define external stakeholders as those actors who do not have an official or contractual link to a project organization but may influence or be influenced by the project (Winch, 2004, p. 323). Concrete examples include, inter alia, communities, end-users, non-governmental organizations (NGOs), and the public. Research and practice have shown that engaging such external stakeholders is a pivotal part of project management (Bayiley and Teklu, 2016; Winch, 2017). By stakeholder engagement, we mean the organizational practices, used to involve external stakeholders in project's activities (cf. Greenwood, 2007), For example, engaging external stakeholders proactively and transparently in projects' operations has been shown to improve project performance,

[1] Aalto University, Finland; University of Oulu, Finland.
[2] University of Oulu, Finland.
* Corresponding author: Jere.Lehtinen@oulu.fi

for instance, by reducing the incidence of protests, appeals, and halts from these stakeholders that would otherwise detrimentally impact projects' operational performance (Aaltonen et al., 2015; Di Maddaloni and Davis, 2017). In addition, stakeholder engagement has been recognized as a central mechanism for enhancing projects' value creation, as it enables a negotiation process with external stakeholders about what value is being created in a project, and how, helping to resolve stakeholder conflicts and create compromises and possible win-win scenarios (Gil and Fu, 2021).

There exist a plethora of practices to engage external stakeholders, such as information seminars and workshops (Lehtinen et al., 2019), but rapid development of social media has offered a completely new channel and means of engaging and communicating with external stakeholders. In the UK, for instance, the London Crossrail and High-Speed Rail 2 project organizations use their official Twitter and Facebook accounts for transparent disclosure of information to their social media communities about upcoming interruptions, decisions, plans, and changing traffic arrangements that might even affect public opinion. In practice, project organizations have, however, adopted various ways of using social media for engagement purposes: while some project organizations utilize social media primarily as a one-way information sharing channel (Ninan et al., 2019; Turkulainen et al., 2015; Williams et al., 2015), others enact more active and proactive strategies and, for example, focus on dialogue to gather feedback and ideas from external stakeholders on project activities (Lehtinen, 2021; Ninan et al., 2020). While information-sharing strategies can inhibit the negotiation process with external stakeholders and creation of value, the active strategies are more resource-intensive and pose other risks, such as counterproductive stakeholder mobilization (cf. Lobo and Abid, 2020). Hence, different approaches to use social media and engage external stakeholders imply different value creation logics and value distribution between project organizations and external stakeholders.

The above nascent evidence of how project organizations use social media and engage external stakeholders is a valuable starting point, but a more systematic understanding of project organizations' roles in and related uses for social media is required to advance theoretical knowledge of how project organizations can utilize social media in external stakeholder engagement.

Our objective in this chapter is to address the above research gap by developing a conceptual framework of the uses of social media in external stakeholder engagement, including a typology of project organizations' roles. In so doing, we focus on the following research questions: *What kinds of roles can project organizations adopt when using social media for external stakeholder engagement? What are the related uses for social media in external stakeholder engagement? What kinds of implications do the different roles have on external stakeholders' engagement, and on value creation and distribution in projects?* In developing the framework (a two-by-two matrix), we draw on instrumental and normative stakeholder theories and approach stakeholder engagement from a managerial viewpoint where a project organization's approach to stakeholder engagement can range from strategic (exclusive approach) to moral (inclusive approach) that represent theoretical extremes and opposite ends of a continuum, forming the first dimension of the framework. In addition, we delve into research on stakeholder communication and adopt the concepts of stakeholder debate and stakeholder dialogue as project organization's two primary modes of communication that also represent theoretical extremes and opposite ends of a continuum, forming the second dimension of our conceptual framework. Thereafter, we discuss the developed framework and analyze the roles that project organizations can adopt (Disengager, Informer, Communicator, Enfranchiser—DICE model) and related uses for social media, including implications on external stakeholders' engagement. We conclude this chapter by first discussing the implications on value creation and distribution in projects, and then suggesting ideas for future research, including potential research questions to be mobilized forward in studying the role of social media in external stakeholder engagement.

9.2 Developing Conceptual Framework: Engagement Approaches and Communication Modes

9.2.1 Stakeholder Engagement

Stakeholder engagement has emerged as the concept defining how various organizations practice stakeholder theory (Kujala and Sachs, 2019), but the concept itself is equivocal and it has been approached from different perspectives and defined in different ways, depending on contextual and empirical considerations (Brodie et al., 2019). At least three general views on stakeholder engagement can be classified: interaction, stakeholder, and managerial views. The interaction view focuses on the interaction between an organization and its stakeholders and seeks to study, e.g., how organizations and stakeholders' actions affect each other (Noland and Phillips, 2010). In turn, the stakeholder view focuses on the agency and perspective of organizational stakeholders, for instance, by scrutinizing stakeholders' experiences with an organization where engagement is often defined as an attribute of stakeholders, e.g., a psychological state or disposition (Brodie et al., 2011). Lastly, the managerial view, being perhaps the most common view in management and business research (*see*, e.g., part III in Harrison et al., 2019), focuses on organizations' agency and management. Maybe the most widely known and utilized definition for engagement is that of Greenwood's (2007, pp. 317–318): 'practices that the organization undertakes to involve stakeholders in a positive manner in organizational activities.' The definition emphasizes morally positive practice that enables collaboration (mutually benefitting relationships) between an organization and its stakeholders, but engagement may also be morally negative practice, masqueraded as corporate social responsibility and used as a deceptive control mechanism (Greenwood, 2007, p. 320). We follow Greenwood's view on stakeholder engagement as it is consistent with our approach to social media as project organizations' managerial practice for engaging external stakeholders.

The morally positive and negative practice of stakeholder engagement draws attention to the ethical side of the concept. The ethical discussion of stakeholder engagement has its roots in Donaldson and Preston's (1995) seminal discussion about the descriptive, instrumental, normative, and managerial aspects of stakeholder theory. Subsequent stakeholder theorists and business ethicists have researched the many aspects of stakeholder theory (*see* Jones and Wicks, 1999; Kaler, 2003; Steurer, 2006) that have also functioned as foundations for the concept and practice of stakeholder management—the predecessor of stakeholder engagement. There are two dominant schools of thought on stakeholder management, based on instrumental and normative stakeholder theories (Goodpaster, 1991). The instrumental view on stakeholder management focuses on single organization's performance outcomes and the instrumental value of stakeholder management, as organizations adopting this view address stakeholders as the means for achieving organizational objectives (ends) (Gibson, 2000). The instrumental view on stakeholder management is also known as the 'management of stakeholders' approach (Savage et al., 1991). In turn, the normative view on stakeholder management sees that stakeholders have intrinsic value, and organizations should consider all stakeholders equally and for their own value, instead of their instrumental value of advancing own organizational goals (Clarkson, 1995). The normative view on stakeholder management is also often called as the 'management for stakeholders' or 'managing for stakeholders' approach (Freeman et al., 2007). The two views are theoretical and extreme cases, and in reality, the case for stakeholder management is something in between because on the one hand, a purely instrumental approach would likely not yield sustainable and long-term outcomes and on the other hand, while stakeholders are of equal value (normative), not all stakeholders could or should be equally involved and considered in all organizational activities (Phillips et al., 2003).

The instrumental and normative stakeholder theories have also had a central role in the development of the concept of stakeholder engagement as research on stakeholder engagement has distinguished between strategic and moral engagement (Noland and Phillips, 2010). Strategic engagement can be seen as the descendent of instrumental stakeholder management, as this approach is inherently focused on achieving strategic purposes and maximizing an organization's own value captured through the engagement of stakeholders (Rasche and Esser, 2006). Strategic engagement can be rationalized, e.g., based on resource-dependency theory, which implies that it is only necessary to engage those stakeholders that possess critical resources for organization's survival (cf. Frooman, 1999). It is worth mentioning that strategic engagement does not imply that organizations would knowingly deceive or harm stakeholders who are disengaged, although this might happen (Noland and Phillips, 2010). Strategic engagement and instrumental stakeholder management can be perceived as synonyms but although they share the same motive and outcome, achievement of strategic organizational goals, the means to achieve the outcome differ. Strategic engagement includes involving stakeholders in organizational activities even though the motive is instrumental, and the involvement can be symbolic with no real opportunities for stakeholders to contribute to organizational activities (Foster and Jonker, 2005). In turn, instrumental stakeholder management does not necessitate the involvement of stakeholders in organizational activities, but stakeholders can be treated superficially and unfairly with other kinds of management activities (Hillman and Keim, 2001; Savage et al., 1991). Stakeholder theorists have argued that strategic engagement is not able to fulfil the paramount proposition of stakeholder theory, which is the purpose of maximizing value creation to stakeholders (Noland and Phillips, 2010).

Moral engagement can be seen as the successor of normative stakeholder management, as this approach is focused on the transparent and genuine engagement of stakeholders in organizational activities and which is seen as the primary purpose of any organization (Reynolds and Yuthas, 2008). Stakeholder theorists have argued that moral engagement can fulfil the main thesis of stakeholder theory: creating value to all stakeholders (Freeman et al., 2007), and more importantly, to sustain the value creation long-term, enabling organizations' long-term survival (Freeman et al., 2010, p. 282). Moral engagement follows the same ideology as normative stakeholder management: considering all stakeholders equally and for their intrinsic value to maximize value creation to all stakeholders. The two can be viewed as synonymous as they share the same motive and outcome, but they can also have slight differences in the way the motive and outcome is fulfilled. Moral engagement, focused on engaging stakeholders, can differ from normative stakeholder management that does not impose engagement as the means to consider stakeholders equally and create value for them, but allows for other ways of interacting with stakeholders.

Moral and strategic engagement represent theoretical extremes. In reality, however, the idea in stakeholder theory and stakeholder engagement for that matter, is not to separate strategic and moral grounds but depict them as a natural amalgamation (Freeman, 1994). But for the purposes of developing a conceptual framework, we adopt these two theoretical extremes as the opposite ends of a continuum. The two approaches to engagement are suitable for understanding the theoretically typical roles of project organization when engaging external stakeholders in social media.

In moral engagement, a project organization adopts an inclusive approach to stakeholder engagement (Di Maddaloni and Davis, 2018; and survival (Eskerod et al., 2015b) and concurrently as many stakeholders as possible in organizational activities transparently and with genuine intentions of maximizing value creation to all stakeholders, including the project organization itself. In the inclusive scenario, a project organization reaches out to several different stakeholders, both external and internal, for achieving a wider breadth of stakeholder engagement (broad engagement of stakeholders (cf. Lehtinen, 2021; Markovic and Bagherzadeh, 2018). In this approach, a project organization also seeks to achieve higher levels of engagement from single stakeholders (Heravi et al., 2015), in which level of engagement is defined as a stakeholder's attribute, a psychological state or disposition (Brodie et al., 2011). Level of engagement is also known as depth of engagement (Lehtinen and Aaltonen, 2020) that relates to breadth and depth of stakeholder relationship (Shymko and Roulet, 2017). In turn, in strategic engagement, a project organization follows an exclusive approach to stakeholder engagement (Di Maddaloni and Davis, 2017; Lehtinen et al., 2019), where it focuses on engaging only a few central stakeholders (e.g., those that possess critical resources and information for the project's progress and survival (Eskerod et al., 2015) and concurrently disengages many other stakeholder groups, for example, by ignoring them or treating them and their concerns superficially. In this exclusive case, the 'breadth of engagement' is narrow as project organization attends only a few stakeholder relationships and even these relationships are primarily approached from an instrumental viewpoint, for advancing project's operational performance (i.e., staying in schedule, budget, and scope) instead of creating value to these stakeholders. Hence, a project organization relying on exclusive approach does not seek to achieve higher levels of engagement from single stakeholders.

The two approaches to stakeholder engagement, inclusive and exclusive, grounded in moral (normative stakeholder theory) and strategic (instrumental stakeholder theory) engagement respectively, form the first

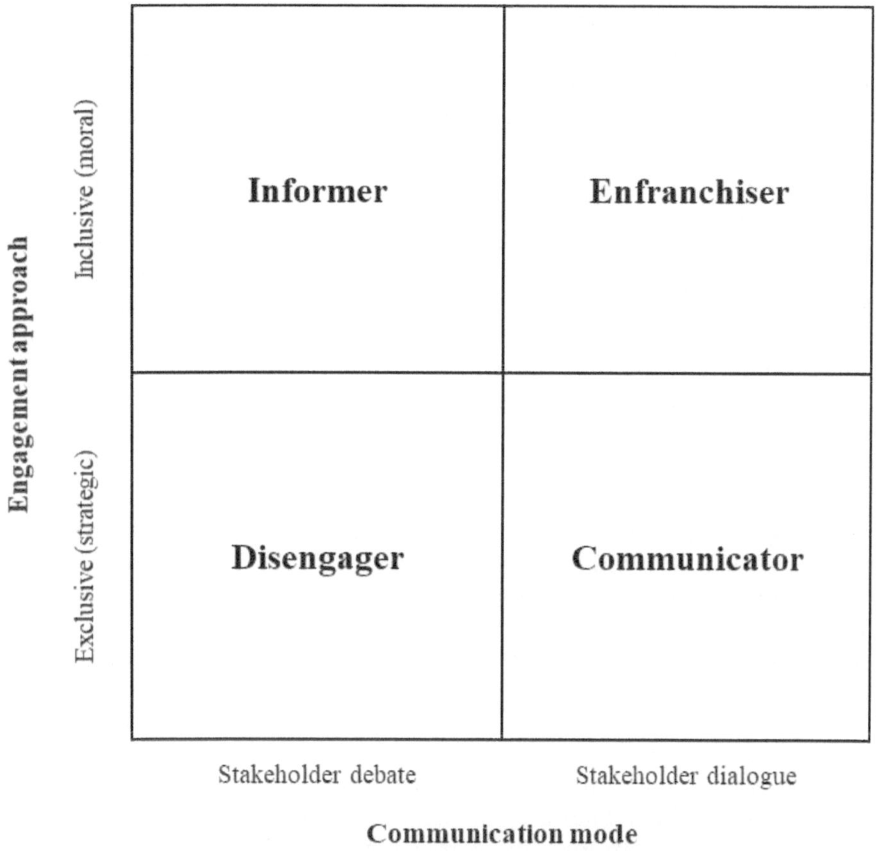

Fig. 9.1: DICE model – conceptual framework of project organizations' roles in using social media for external stakeholder engagement.
(Alt Text: Conceptual framework in a matrix form covering communication mode and engagement approach)

dimension of the conceptual framework (*see* Fig. 9.1 vertical axis). The breadth of engagement is wide in the inclusive approach and narrow in the exclusive approach.

9.2.2 Stakeholder Communication

Stakeholder engagement is primarily enacted through communication (Freeman et al., 2017, pp. 7-8). Stakeholder communication can be defined as indirect and direct conversations between an organization and its stakeholders (Kujala and Sachs, 2019). From the perspective of stakeholder communication, the different organizational practices used to engage stakeholders in organizational activities can be considered as communication media where an organization and its stakeholders interact.

Stakeholder communication has a central purpose in projects as direct and indirect conversations enable understanding stakeholders' distinct requirements (Butt et al., 2016) that can be taken into account in projects' activities and decision-making to ensure value creation and achievement of operational performance (Maylor and Turner, 2017). But, due to the temporal nature of projects, there exist limited windows for communication, challenging timely interaction (Burke and Morley, 2016). The temporariness also implies that stakeholders and their needs can change over time, which brings uncertainty into stakeholder communication and engagement (Maylor and Turner, 2017). Stakeholder research has argued that constant and proactive communication with stakeholders may be necessary for engaging stakeholders successfully (Passetti et al., 2019), especially in the context of projects (Derakhshan et al., 2019; Unterhitzenberger et al., 2020). However, this is difficult to achieve since resources for communication are limited in projects both in time and quantity (Di Maddaloni and Davis, 2018).

Stakeholder communication is often approached from a contingency view, where an organization devises its communication media, depending on the uncertainty and equivocality of information that needs to be processed to various stakeholders (Daft and Lengel, 1986). The information processing requirements stemming from uncertainty and equivocality define the richness and amount of information that an organization can process to its stakeholders. In projects, this implies that project organizations need to balance the breadth (i.e., extent to which an organization communicates with multiple stakeholders) and depth (i.e., intensity with which an organization communicates with a single stakeholder) of communication when planning communication media to ensure that there is a fit between information processing capabilities and requirements (Monteiro de Carvalho, 2013). For example, traditional communication media, including, e.g., media outreaches, webpages, bulletins, press releases, leaflets, and articles are appropriate for achieving the breadth of communication, because these are suitable for disseminating information to a broad audience of stakeholders (Gil, 2010). However, these channels lack interaction with stakeholders, making it difficult to achieve the depth of communication with stakeholders; that is, these media facilitate communication to stakeholders, but are not appropriate for communicating with stakeholders. In turn, more interactive media, such as workshops and information events and seminars, facilitate interaction and communication with stakeholders, helping to achieve depth of communication with specific stakeholder groups (Lehtinen et al., 2019). But the downside often is that the number of different stakeholders that can process the information is lower in these physical events, implying that these interactive media may fail to achieve wider breadth of communication as compared to traditional media. Nevertheless, interactive media facilitates both communication to and with organizational stakeholders.

Communication to and with stakeholders relate to two communication modes that organizations can adopt (Freeman et al., 2017). Communication to stakeholders is also known as stakeholder debate where an organization tries to influence stakeholders and their behaviors through one-way communication (Lehtimaki and Kujala, 2017). Organizations opting for this mode can be driven by instrumental value of stakeholder communication as they defend their agenda to stakeholders and see stakeholders as barriers and competitors for achieving own organizational goals (Kaptein and Van Tulder, 2003; Kujala and Sachs, 2019). This communication mode goes hand in hand with traditional communication media that are appropriate for convincing and defending organizational agendas, often executed through formal communication content. In turn, communication with stakeholders is also known as stakeholder dialogue in which organizations focus on empathic collaboration through two-way communication and by giving a voice to stakeholders and genuinely listening to their concerns (Lehtimaki and Kujala, 2017). Organizations opting for this mode are driven by the normative grounding of stakeholder communication as they seek to exchange insights, opinions, and ideas constructively to co-create value with stakeholders (Kaptein and Van Tulder, 2003). Stakeholder dialogue approach is well-suited with interactive media because they enable two-way communication, informal content, and sharing of viewpoints for building deeper relationships with stakeholders (Aaltonen and Turkulainen, 2018). Stakeholder theory has in general acknowledged the strengths of stakeholder dialogue in engaging and creating value with stakeholders (Bebbington et al., 2007), but there are also downsides due to high resource requirements, which are a challenge especially in projects that possess very scarce resources for stakeholder communication and engagement (Di Maddaloni and Davis, 2018).

As discussed in introduction, project organizations are increasingly using social media for stakeholder engagement and communication purposes. Social media include computer-mediated technologies and applications like Twitter, Facebook, YouTube, and Instagram that enable the creation and sharing of textual, audio- and video-based information in virtual networks and communities (Kietzmann et al., 2011). These media enable project organizations to adopt both stakeholder dialogue and debate modes, facilitating achievement of both depth and breadth of stakeholder communication. For example, social media enable rapid information distribution about project activities and also dialogues with external stakeholders to exchange knowledge, thus offering an efficient and effective way of reaching a wide range of stakeholders (Lovejoy and

Saxton, 2012). The functionalities of these media may facilitate achieving both breadth of stakeholder engagement and higher levels of engagement from single stakeholders.

Despite the potential of stakeholder dialogue and depth of communication, empirical research has shown that project organizations have thus far utilized social media primarily for one-way communication and achieving the breadth of communication. For instance, some project organizations have used Instagram for providing information to stakeholders, Facebook for creating a sense of community among stakeholders, and YouTube for advertising project activities to stakeholders (Turkulainen et al., 2015). Rather similarly, some project organizations have offered progress updates, entertainment, and information on sustainability issues (Lehtinen, 2021), built legitimacy for themselves and their activities, delivered targeted marketing and advertisements, and appealed to stakeholders to foster a positive project image and support for project activities (Ninan et al., 2019, 2020). Also, some project organizations have recently used mobile social media (WeChat and WhatsApp applications) for disclosing information internally (Zhang et al., 2018). In addition to direct communication purposes, some research suggests that social media is useful for analysis purposes to understand external stakeholders and their online behavior facilitating enactment of subsequent communication to/with them (Lobo and Abid, 2020; Williams et al., 2015). There is also some evidence of project organizations using social media for two-way communication purposes (Lehtinen, 2021; Ninan et al., 2020), but the above examples show that project organizations have still a lot to learn about how to unlock the full potential of social media in stakeholder communication and engagement.

The two communication modes—stakeholder debate and dialogue—represent theoretical extremes, and it is likely that organizations in reality utilize both modes but to varying degrees and purposes. For purposes of developing the conceptual framework, we adopt these two communication modes as the opposite ends of a continuum (Fig. 9.1, horizontal axis). The two modes provide appropriate framework for understanding the theoretically typical roles of project organization in using social media for stakeholder communication.

9.3 Framework Analysis: Project Organization's Roles and Related Uses for Social Media

We named the conceptual framework as DICE model (summarized in Fig. 9.1) based on the four derived roles for project organizations (Disengager, Informer, Communicator, Enfranchiser) using social media for external stakeholder engagement. Next, in Subsections 9.3.1–9.3.4, we will explain these roles and discuss the related uses for social media and implications on external stakeholders' engagement and on value creation and distribution in projects. On a side note, the framework is general in the sense that it could also be used for understanding the role of other organizational practices than social media in stakeholder engagement, also in other contexts than projects or external stakeholders.

9.3.1 Disengager

Project organizations, which are characterized as disengagers, adopt an exclusive approach to stakeholder engagement and rely on stakeholder debate as their primary communication mode when using social media for external stakeholder engagement. This role overall comprises a very passive and unorganized usage mode to engaging external stakeholders as in practice, project organization disengages external stakeholders. The previous means that a project organization may have opted to use a social media application for external stakeholder engagement but does not use it systematically but instead very inconsistently, withholding information that results in narrow breadth of engagement. For example, posting only general announcements occasionally (i.e., no transparent information provided but output is very limited and controlled) and not replying to external stakeholders' comments or questions at all or replying either in a very superficial manner to shut down any constructive discussion. Shutting down two-way communication likely leads to stakeholders' lower levels of engagement as well. In this case, the potential of social media is not realized as social media is only a symbolic engagement practice and plays no real role in external stakeholder engagement. Project organizations adopting disengager role perceive external stakeholders as holdouts and risks that are barriers to value creation and project success (cf. Liu et al., 2018).

In this role, project organizations use social media to safeguard the project from external disturbances (cf. Lundin and Söderholm, 1995) based on the rationale that any changes to project activities would be harmful and costly in achieving operational performance (Flyvbjerg et al., 2003). That is, social media is not used to create value for stakeholders but only to secure value capture for project organization. However, the passiveness likely backfires as external stakeholders experience that they and their concerns are not addressed in any manner, resulting in strong opposition (Lobo and Abid, 2020) and even social media riots that can severely hamper the project's public image. Also, there might be cascading effects as external stakeholders may start rioting and appealing through other media to draw attention to their issues, which can result in serious setbacks, such as legal disputes, litigation, unplanned and unwanted project changes, halts, and loss of reputation (Aaltonen et al., 2015; Di Maddaloni and Davis, 2017).

In a disengager role, there is a poor or no fit between the information-processing requirements and capabilities. External stakeholders and project organizations often have conflicting goals and interests, meaning that the two actors are very different, but the two are still very interdependent, because they need to work together toward the shared and unique project objective, limited resources, and time (Jones and Lichtenstein, 2008). The previous high difference and high interdependency indicates that there is a need to process large amounts of information in a rich format to reduce equivocality and uncertainty (cf. Daft and Lengel, 1986). But in a disengager role, a project organization fails to do the previous and does not process the required amount and richness of information to external stakeholders via the social media, thus failing to reduce equivocality and uncertainty in project activities that can further lead to problems in project implementation. In addition, a project organization may not analyze the external environment properly but neglects it, resulting in incorrect interpretations about the information-processing requirements, further contributing to the poor fit between requirements and capabilities.

9.3.2 Informer

Project organizations, which adopt an informer's role, rely on an inclusive approach to stakeholder engagement but use stakeholder debate as the primary communication mode in social media. This approach includes an active and organized usage mode of communicating to external stakeholders, meaning that the use of social media is a systematic engagement practice, and it is used consistently and with consideration, resulting in wide breadth of engagement. This approach manifests the prevailing roles that project organizations have hitherto adopted. As discussed previously, empirical evidence has shown how project organizations have used social media applications for information distribution, entertainment, and advertising and marketing to improve and maintain projects' and project organizations' reputation (Lehtinen, 2021; Ninan et al., 2019, 2020; Turkulainen et al., 2015). In this role, the potential of social media is partly realized, because social media is adopted as a real engagement practice and contributes to value creation by keeping external stakeholders informed and updated. However, project organizations adopting this role do not stimulate dialogue, ask for feedback, or offer external stakeholders with authentic contributor roles and access to influence the project (e.g., opportunities to advance their ideas, visions, social needs, and values), meaning that part of the potential of social media is not realized. The previous means that external stakeholders' level of engagement is likely not very high. Still, informers perceive external stakeholders in a positive light and that engaging them is important for project's value creation and success.

Informers are similar to disengagers in that they do not offer external stakeholders with opportunities to influence project activities because they do not opt for two-way communication via stakeholder dialogue. Hence, project organizations use social media similarly to safeguard the project from external stakeholders' pressures as any unplanned or unwanted changes are experienced as deviations that can decrease the project performance. Offering external stakeholders with real opportunities to influence is seen as a risk because it might trigger undesirable stakeholder demands, challenging control over the project (cf. Gond et al., 2016). Thus, social media is used to create value but not to co-create value with external stakeholders, because value is being created according to project organizations' terms. Hence, some value, like informational value, is created for external stakeholders, but majority of the value is distributed to project organization. The benefit of the informer role over disengager role is that it will less likely mobilize external stakeholders to oppose the project but keeps them dormant and satisfied (cf. Lobo and Abid, 2020). But some external stakeholders are still likely to resist the project because they are not offered with opportunities to influence project operations and what value is being created, and how.

There is a partial fit between the information processing requirements and capabilities when adopting an informer role. Informers are able to process large amounts of information to reduce uncertainty as they focus on engaging external stakeholders through constant information distribution. But due to the lack of two-way communication, they fail to always reduce equivocality because one-way communication does not always offer rich enough information to satisfy external stakeholders' needs. That is, whenever external stakeholders request for more information by asking questions or try to influence project activities via social media, informers tend to ignore these requests and external stakeholders are left confused and it is likely that they misinterpret some information. Thus, a project organization succeeds in processing the required amount of information to reduce uncertainty, but fails to process the required richness of information, thus failing to reduce equivocality. The previous implies that informers analyze external environment partially and make both correct (need to disseminate information) and incorrect (no need to open dialogue or opportunities to influence) deductions about the information processing requirements.

9.3.3 Communicator

Communicators rely on an exclusive approach to stakeholder engagement but use stakeholder dialogue as the primary communication mode in social media. Like disengagers, communicators are passive in communication, but they have adopted social media as a systematic practice for external stakeholder engagement and use it consistently and with careful consideration. However, the focus in this role is to maximize project organizations' value capture over broader value creation as the intentions of using social media are very strategic. In this role, project organizations adopt an expert role and critically control the interaction and communication by opening specific issues for discussion and reply very selectively to external stakeholders' questions and concerns in social media, resulting in narrow breadth of engagement. The previous implies that despite relying on stakeholder dialogue, communicators are still quite monologists, highlighting organizations' traditional communication roles (cf. Passetti et al., 2019). There is some empirical evidence that suggests the previous, as one case study found that a project organization practiced its role as an expert, opening discussion around pre-selected issues and replied selectively to external stakeholders' messages and queries, overall governing the communication and interaction carefully to define a more closed community that had limited contribution opportunities (Lehtinen, 2021). In practice, external stakeholders are provided with only symbolic roles and their opportunity to influence pre-determined issues is merely ostensible. The potential of social media is realized partly because social media is adopted as a systematic engagement practice, offering some information to external stakeholders, but due to its symbolic role in external stakeholder engagement, it does not contribute genuinely to value creation. In general, communicators perceive external stakeholders important through their instrumental value in advancing project's goals.

Communicators are similar to disengagers in the way that they do not seek to engage external stakeholders genuinely, but unlike disengagers, they have opted for two-way communication via stakeholder dialogue. However, stakeholder dialogue is used to control the value creation process and secure benefits for project organization, again relating to safeguarding project from external stakeholders' views, thereby ensuring that no unwanted or unplanned changes occur. Hence, social media is used to capture value instead of creating value, because value is being created and captured only on project organizations' terms. However, external stakeholders are offered with opportunities to influence (although symbolic), implying that they are kept partly satisfied and dormant, likely resulting in less opposition. Some external stakeholders might still resist strongly because the agenda is set by project organizations which are possibly in better positions to influence the communication and what is considered appropriate and legitimate, relating to power dynamics and politics in stakeholder interaction (Chow and Leiringer, 2020, 2021; Mitchell et al., 1997). Also, external stakeholders may well experience that their viewpoints are considered only marginally as specific pre-determined opportunities to influence project's activities and decision-making are mechanisms that reflect project organizations' dominant business interests with the purpose of achieving control and value capture (cf. Spence and Rinaldi, 2014). It is still worth mentioning that communicators create some value to external stakeholders, because they provide informational value through communication and despite the opportunities to influence being symbolic, these still create some value as external stakeholders have at least a small opportunity to influence what value is being created and how. The previous implies that some external stakeholders' level of engagement may even be higher.

There is a partial fit between the information-processing requirements and capabilities when adopting a communicator role. Communicators can process rich information and reduce equivocality through two-way communication (although selectively) but due to the strategic orientation and targeted information sharing, they do not offer large amounts of information and reduce uncertainty in most cases, especially if compared to informers who broadcast information constantly. In other words, communicators address external stakeholders' concerns and questions, although with self-serving purposes and selectively, and offer opportunities to influence pre-determined issues. Therefore, communicators succeed in processing the required richness of information to reduce equivocality of project activities but fail to process the required amount of information most often, thus failing to reduce uncertainty. Like informers, communicators analyze external environment partially correct and make both correct (need to open dialogues and offer opportunities to influence) and incorrect (no need to dispense information extensively) interpretations about the information processing requirements.

9.3.4 Enfranchiser

Enfranchisers rely on an inclusive approach to stakeholder engagement and use stakeholder dialogue as the primary communication mode in social media. Enfranchisers play the most active and organized role in communication as social media has been adopted as a systematic practice for co-creating value with external stakeholders, resulting in both wide breadth of engagement and stakeholders' high levels of engagement. The previous indicates that enfranchisers utilize the full potential of social media. In this context, value co-creation means that project organizations enfranchise external stakeholders by granting them both decision-making and claimancy rights in projects (Gil and Fu, 2021; Klein et al., 2019). External stakeholders' decision-making rights are naturally not equal to project organizations which have more decision-making authority due to their position in projects' organizational design (cf. Klein et al., 2019).

However, the above still means that project organizations offer external stakeholders real contributor roles and opportunities to influence project activities and thus to create value together with project organizations. In turn, claimancy rights in this context establish that external stakeholders can capture their fair share of the jointly-created value. In general, enfranchisers see external stakeholders as enablers of value creation and success (cf. Garriga, 2014).

Enfranchisers have a dedicated social media, communication or PR manager who seeks to find win-win situations together with external stakeholders, instead of compromises or win-lose situations (Lehtinen, 2021, p. 120). For example, project organizations adopting this role use the interactive features of social media to offer external stakeholders with contributor roles to work collaboratively around issues that are relevant for external stakeholders, such as social, financial, and environmental issues (cf. Cho et al., 2014; Lee et al., 2013). The previous also means that external stakeholders are asked to generate content and cast votes concerning project issues through social media reactions and comments (Lehtinen, 2021). Enfranchisers utilize all the functionalities of social media, including cutting-edge, multi-media visualizations, and other emotional and logical appeals (cf. Ashley and Tuten, 2015; Viglia et al., 2018). Enfranchisers also systematically stimulate dialogue in general with and among external stakeholders by inviting questions and feedback, encouraging active participation and mobilizing them to dynamically co-develop novel ideas and solutions that can even facilitate co-creating innovations, being a relevant value creation mechanism (cf. Martini et al., 2014). The above uses highly contribute to external stakeholders' higher levels of engagement. Enfranchisers may also be directly in touch with external stakeholders' accounts, groups, and communities instead of just creating content on their own social media account and pages, further facilitating reaching even wider breadth of engagement.

In the case of enfranchiser, there is a perfect fit between the information-processing requirements and capabilities. Enfranchisers process rich information through constant dialogue and thus reduce equivocality of project activities, and due to the inclusive approach to stakeholder engagement, they also disseminate transparent information constantly, thus reducing uncertainty in project activities. Enfranchisers invest a lot of resources in social media activities and are thus active in analyzing the external environment, facilitating making correct interpretations about the information-processing requirements. Naturally, one main downside when adopting an enfranchiser role is that it requires high resource investments at least initially to learn the ropes, which is a challenge in projects that are unique and have limited resources for stakeholder communication and engagement to begin with (Di Maddaloni and Davis, 2018). Adopting an enfranchiser role successfully may thus require prioritizing the role of stakeholder engagement and communication in project organizing in the first place and gathering learning of best practices. Another issue relates to the openness dictated by the enfranchiser role, that is, enfranchisers offer transparent information continuously and external stakeholders are empowered to influence projects, which means that external stakeholders are aware of projects' issues (Liu et al., 2018) and likely exercise their power autonomously through various activities that influence projects (Lobo and Abid, 2020; Nguyen et al., 2019). So, the trade-off then can be that while external stakeholders create value with enfranchisers through influencing activities that cause changes and improvements in projects (e.g., overall, more valuable project outcome), they can also incur additional costs and/or schedule amendments that can lead to the project organization capturing a smaller portion of the overall value (e.g., due to decreased project performance). The two issues above may be some key reasons for project organizations refraining from adopting the enfranchiser role widely in projects.

9.4 Discussion

The implications of the four different roles on the use of social media and value creation and distribution in projects are summarized in Fig. 9.2.

DICE model suggests a typology of project organization's roles in using social media for external stakeholder engagement, including the roles of disengager, informer, communicator, and enfranchiser. It is worth mentioning that these roles are extreme examples/ideal cases based on the theoretical framework and the roles are likely to be different in the real world (e.g., combinations of these and other roles). The analysis illustrates how the four roles are connected to three different value orientations in projects: value capture, creation, and co-creation. Adopting different value orientations fundamentally affects the way project organizations utilize social media in practice for external stakeholder engagement.

In the case of value capture, project organizations adopt either a disengager or communicator role and use social media primarily strategically to ensure own value capture and attainment of operational project performance. In value capture orientation, external stakeholders are seen as risks, and they play primarily an instrumental role as social media is used more for its symbolic role. In this scenario, stakeholder debate is used to defend project organization's agenda and activities in a way that shuts down any constructive discussion with/to external stakeholders. Little to no value is created for external stakeholders as value is distributed to project organization side primarily. The previous

Engagement approach	Stakeholder debate	Stakeholder dialogue
Inclusive (moral)	**INFORMER** • Active and organized usage mode • Value creation through one-way and transparent information dissemination • External stakeholders important to project's success and value creation • No value co-creation: external stakeholders lack real contributor roles • Value is distributed unevenly • Wide breadth but low level of engagement • Partial fit between information processing requirements and capabilities	**ENFRANCHISER** • Active and organized usage mode • Dedicated social media manager • External stakeholders co-create value with project organizations • Enfranchisement: decision-making and claimancy rights granted for externals • Win-win scenarios via active and transparent dialogue and information dissemination • Wide breadth and high level of engagement • Perfect fit between information processing requirements and capabilities
Exclusive (strategic)	**DISENGAGER** • Passive and disorganized usage mode • External stakeholders are barriers to value • Focus on project organization's value capture over broader value creation • Information dissemination is inconsistent and not transparent • Social media has a symbolic role in external stakeholder engagement • Narrow breadth and low level of engagement • No fit between information processing requirements and capabilities	**COMMUNICATOR** • Passive and organized usage mode • External stakeholders are important due to their instrumental value • Externals are offered symbolic contributor roles through dialogue • Little information is disseminated to control carefully value creation process • Focus on project organization's value capture over broader value creation • Narrow breadth but high level of engagement • Partial fit between information processing requirements and capabilities

Communication mode

Fig. 9.2: Implications of the four roles on the use of social media, external stakeholders' engagement, and value creation and distribution in projects.
(Alt Text: Implications of the four roles discussed in a matrix with communication mode on the X-axis and engagement approach on the Y-axis)

implies a closed approach where external stakeholders have only a passive role of receiving information and there are no opportunities to interact with project organizations and influence value creation processes. Project organizations can also use stakeholder dialogue through a symbolic approach, in which external stakeholders are offered only ostensible roles to influence project activities and project organizations strictly govern value creation processes for their own benefit. Still, some informational value is created to external stakeholders through communication and despite the opportunities to influence being symbolic, these can still create some value to external stakeholders because they have a small voice in influencing what value is being created and how.

In turn, project organizations can focus on value creation by adopting an informer role, where social media is used for constant and transparent information dissemination. In this orientation, project organizations recognize external stakeholders' importance for project success and thus adopt an inclusive approach. However, project organizations recognize that open stakeholder dialogue is too risky for projects' progress as open interaction might trigger unwanted deviations, and thus, instead, they rely on stakeholder debate as the primary communication mode. Nevertheless, this approach creates constant, informational value for external stakeholders and can achieve wide breadth of engagement, being the most prevalent role that project organizations have so far adopted when using social media. However, external stakeholders' level of engagement is likely low in this orientation as there is no room for open interaction through dialogue.

Lastly, project organizations can focus on value co-creation, in which social media is used for enfranchising external stakeholders with decision-making and claimancy rights. In this orientation, external stakeholders are seen as partners and contributors for value creation, and they are engaged transparently through constant dialogue. Project organizations adopting this orientation, focus on creating win-win scenarios by inviting questions and feedback, encouraging active participation, and mobilizing external stakeholders to dynamically co-develop novel ideas and solutions, that can even facilitate co-creating innovations, thus being a relevant value-creation mechanism. Value co-creation orientation focuses on utilizing the full potential of social media in external stakeholder engagement that helps achieve a wide breadth of engagement as well as increase external stakeholders' level of engagement. However, as discussed in the previous section, adopting this value co-creation orientation is difficult in practice as it has its own share of downsides and overcoming these may require a fundamentally different attitude towards external stakeholders and stakeholder engagement.

Although empirical evidence so far has suggested that project organizations have not unlocked the full value creation potential of social media in external stakeholder engagement, the framework's connection between value orientations and project organizations' roles in social media highlight how social media can enable true enfranchisement of external stakeholders and value co-creation in projects. Social media can transform the dyadic logic of external stakeholder engagement and value creation processes toward a networked, multi-actor logic. Instead of building dyadic relationships with external stakeholders in traditional media, social media enable real-time interaction with a vast network of external stakeholders in rich audio-visual format that is not possible to achieve in traditional media. Social media not only enables timely information dissemination to a broad network of external stakeholders with just a few resource investments (cost efficiency), but it also introduces functionalities that enable joint problem-solving, crowdsourcing, and co-development of solutions, ideas, and innovations rapidly that can be initiated by both external stakeholders and project organizations in an equal manner. That is, social media can challenge and transform the traditional organizational design and roles of project actors through enfranchisement in a way that enables value co-creation processes in a more boundaryless multi-actor environment.

9.5 Conclusion

The purpose of this chapter was to develop a better understanding of how project organizations can use social media for engaging external stakeholders. In so doing, we developed a conceptual framework of the uses of social media in external stakeholder engagement, including a typology of project organizations' roles. We then analyzed the implications of the framework for external stakeholders' engagement, and for value creation and distribution in projects. We believe that the framework (DICE model) and conceptual advancements brought forward in this chapter are valuable to project management research, as earlier research and empirical evidence have offered so far only illustrative examples of project organizations' roles and the different ways how social media can be used in projects to engage external stakeholders (cf. Lehtinen, 2021; Ninan et al., 2020, 2019; Turkulainen et al., 2015; Williams et al., 2015; Zhang et al., 2018). Indeed, our developed framework offers a more systematic and theoretical examination of the uses and roles, providing explanatory power as it can guide thinking and offer ideas about the factors (engagement approach and communication mode) that affect project organizations' social media practices, external stakeholders' engagement, and value creation and distribution in projects. The developed DICE model is among the first to scrutinize these previous issues, offering new insights of the role of social media in external stakeholder engagement.

The developed framework is a valuable starting point, but further empirical and conceptual research is required at least on the following issues to better understand the role of social media in external stakeholder engagement and value creation.

Context and other factors play a key role in understanding, e.g., the boundary conditions, role of time, other (causal) mechanisms, and contextual idiosyncrasies of the developed DICE model. In the following, we provide some preliminary research questions related to these issues. What other factors than engagement approach and communication mode influence project organizations' social media roles and value creation and distribution in projects, and how? How does project context (e.g., infrastructure, industrial, IT, healthcare, organizational change, and, public, private, third sector) influence project organizations' social media roles and practices in projects? How does the type of social media (e.g., Twitter, Facebook, YouTube, Instagram) influence project organizations' social media behavior and value creation and distribution in projects? How do project organizations shift from one social media role to the other over project life cycle (e.g., what is the role of time/project phases, what kinds of transition dynamics exist, and what factors contribute to role changes over time)?

Value creation process: In this research, we focused on external stakeholder engagement in social media and its implications for value creation and distribution in projects from stakeholder theory perspective. But, shifting the focus on value creation itself is likely to yield new insights of how external stakeholders and stakeholder engagement function as mechanisms in value creation and particularly what value is in this context and how it is created. The following questions are examples of potentially relevant research questions. How do project organizations and external stakeholders co-create value through social media in practice (e.g., through what kinds of mechanisms and who orchestrates the different mechanisms and overall process)? What kind of value is created in social media (e.g., mere informational value related to the project vs. value related to changes/improvements in actual project outcomes/end-product)? How is value distributed in social media (e.g., who has claimancy rights, over what value, and how)? How do the different value orientations (e.g., value capture, value creation, value co-creation) come into being in practice and how do they evolve over project life cycle? How does external stakeholders' enfranchisement occur in social media (e.g., through formal agreements vs. informal arrangements, and who initiates and governs such agreements and arrangements)?

External stakeholders' role: We adopted a managerial perspective in this study and focused on project organizations' agency in social media. However, focusing on external stakeholders' viewpoint and the interaction between external stakeholders and project organizations can facilitate developing a more detailed understanding of both external stakeholder engagement and value creation in social media. The study by Lobo and Abid (2020) is a valuable starting point on external stakeholder mobilization in social media and the following build some additional research questions that we consider relevant in future research. What kinds of roles do external stakeholders have in social media, and how do they contribute to value (co-)creation in projects? How do external stakeholders shift from one social media role to the other over project life cycle (e.g., what is the role of time/project phases, what kinds of transition dynamics exist, and what factors contribute to role changes over time)? How do external stakeholders interact with project organizations in social media over time, and why (e.g. what kinds of activities do they devise and use in different project phases, and how)? How are external stakeholders engaged (effectively) in project activities via social media, and what does engagement mean in this setting?

References

Aaltonen, K., Kujala, J., Havela, L., and Savage, G. (2015). Stakeholder dynamics during the project front-end: The case of nuclear waste repository projects, *Project Management Journal*, **46**(6): 15–41. https://doi.org/10.1002/pmj.21549.

Aaltonen, K., and Turkulainen, V. (2018). Creating relational capital through socialisation in project alliances, *International Journal of Operations & Production Management*, **38**(6): 1387–1421. https://doi.org/10.1108/IJOPM-02-2017-0091.

Ashley, C., and Tuten, T. (2015). Creative strategies in social media marketing: An exploratory study of branded social content and consumer engagement, *Psychology & Marketing*, **32**(1): 15–27. https://doi.org/10.1002/mar.20761.

Bayiley, Y.T., and Teklu, G.K. (2016). Success factors and criteria in the management of international development projects, *International Journal of Managing Projects in Business*, **9**(3): 562–582. https://doi.org/10.1108/IJMPB-06-2015-0046.

Bebbington, J., Brown, J., Frame, B., and Thomson, I. (2007). Theorising engagement: The potential of a critical dialogic approach, *Accounting, Auditing and Accountability Journal*, **20**(3): 356–381. https://doi.org/10.1108/09513570710748544.

Brodie, R.J., Hollebeek, L.D., Jurić, B., and Ilić, A. (2011). Customer engagement: Conceptual domain, fundamental propositions, and implications for research, *Journal of Service Research*, **14**(3): 252–271. https://doi.org/10.1177/1094670511411703.

Brodie, R.J., Fehrer, J.A., Jaakkola, E., and Conduit, J. (2019). Actor engagement in networks: Defining the conceptual domain, *Journal of Service Research*, **22**(2): 173–188. https://doi.org/10.1177/1094670519827385.

Burke, C.M., and Morley, M.J. (2016). On temporary organisations: A review, synthesis and research agenda, *Human Relations*, **69**(6): 1235–1258. https://doi.org/10.1177/0018726715610809.

Butt, A., Naaranoja, M., and Savolainen, J. (2016). Project change stakeholder communication, *International Journal of Project Management*, **34**(8): 1579–1595. https://doi.org/10.1016/j.ijproman.2016.08.010.

Cho, M., Schweickart, T., and Haase, A. (2014). Public engagement with non-profit organisations on Facebook, *Public Relations Review*, **40**(3): 565–567. https://doi.org/10.1016/j.pubrev.2014.01.008.

Chow, V., and Leiringer, R. (2020). The practice of public engagement on projects: From managing external stakeholders to facilitating active contributors, *Project Management Journal*, **51**(1): 24–37. https://doi.org/10.1177/8756972819878346.

Chow, V., and Leiringer, R. (2021). Public engagement events and the management of external stakeholders: Artifacts as boundary objects or tools of discipline and control? *Project Management Journal*, **52**(1): 61–74. https://doi.org/10.1177/8756972820940475.

Clarkson, M.B.E. (1995). A stakeholder framework for analyzing and evaluating corporate social performance, *The Academy of Management Review*, **20**(1): 92–117.

Daft, R.L., and Lengel, R.H. (1986). Organizational information requirements, media richness and structural design, *Management Science*, **32**(5): 554–571.

Derakhshan, R., Turner, R., and Mancini, M. (2019). Project governance and stakeholders: A literature review, *International Journal of Project Management*, **37**(1): 98–116. https://doi.org/10.1016/j.ijproman.2018.10.007.

Di Maddaloni, F., and Davis, K. (2017). The influence of local community stakeholders in megaprojects: Rethinking their inclusiveness to improve project performance, *International Journal of Project Management*, **35**(8): 1537–1556. https://doi.org/10.1016/j.ijproman.2017.08.011.

Di Maddaloni, F., and Davis, K. (2018). Project manager's perception of the local communities' stakeholder in megaprojects. An empirical investigation in the UK, *International Journal of Project Management*, **36**(3): 542–565. https://doi.org/10.1016/j.ijproman.2017.11.003.

Donaldson, T., and Preston, L.E. (1995). The stakeholder theory of the corporation: Concepts, evidence, and implications, *Academy of Management Review*, **20**(1): 65–91. https://doi.org/10.5465/amr.1995.9503271992.

Eskerod, P., Huemann, M., and Ringhofer, C. (2015a). Stakeholder inclusiveness: Enriching project management with general stakeholder theory, *Project Management Journal*, **46**(6): 42–53. https://doi.org/10.1002/pmj.21546.

Eskerod, P., Huemann, M., and Savage, G. (2015b). Project stakeholder management—past and present, *Project Management Journal*, **46**(6): 6–14. https://doi.org/10.1002/pmj.21555.

Flyvbjerg, B., Skamris, M.K., and Buhl, S.L. (2003). How common and how large are cost overruns in transport infrastructure projects? *Transport Reviews*, **23**(1): 71–88. https://doi.org/10.1080/01441640309904.

Foster, D., and Jonker, J. (2005). Stakeholder relationships: The dialogue of engagement, *Corporate Governance*, **5**(5): 51–57. https://doi.org/10.1108/14720700510630059.

Freeman, R.E. (1994). The politics of stakeholder theory: Some future directions, *Business Ethics Quarterly*, **4**(4): 409–421. https://doi.org/10.2307/3857340.

Freeman, R.E., Harrison, J.S., and Wicks, A.C. (2007). *Managing for Stakeholders: Survival, Reputation, and Success*, Yale University Press.

Freeman, R.E., Harrison, J.S., Wicks, A.C., Parmar, B.L., and De Colle, S. (2010). *Stakeholder Theory: The State of the Art*, Cambridge University Press.

Freeman, R.E., Kujala, J., and Sachs, S. (2017). Stakeholder Engagement: Clinical Research Cases, Springer International Publishing. https://doi.org/10.1007/978-3-319-62785-4_8.

Frooman, J. (1999). Stakeholder influence strategies, *The Academy of Management Review*, **24**(2): 191–205.

Garriga, E. (2014). Beyond stakeholder utility function: Stakeholder capability in the value creation process, *Journal of Business Ethics*, **120**(4): 489–507. https://doi.org/10.1007/s10551-013-2001-y.

Gibson, K. (2000). The moral basis of stakeholder theory, *Journal of Business Ethics*, **26**(3): 245–257. https://doi.org/10.1023/a:1006110106408.

Gil, N.A. (2010). Language as a resource in project management: A case study and a conceptual framework, *IEEE Transactions on Engineering Management*, **57**(3): 450–462. https://doi.org/10.1109/TEM.2009.2028327.

Gil, N.A., and Fu, Y. (2021). *Megaproject Performance, Value Creation and Value Distribution: An Organizational Governance Perspective*, Academy of Management Discoveries. https://doi.org/10.5465/amd.2020.0029.

Gond, J.P., Barin Cruz, L., Raufflet, E., and Charron, M. (2016). To frack or not to frack? The interaction of justification and power in a sustainability controversy, *Journal of Management Studies*, **53**(3): 330–363. https://doi.org/10.1111/joms.12166.

Goodpaster, K.E. (1991). Business ethics and stakeholder analysis, *Business Ethics Quarterly*, **1**(1): 53–73. https://doi.org/10.2307/3857592.

Greenwood, M. (2007). Stakeholder engagement: Beyond the myth of corporate responsibility, *Journal of Business Ethics*, **74**(4): 315–327. https://doi.org/10.1007/s10551-007-9509-y.

Harrison, J.S., Barney, J.B., Freeman, R.E., and Phillips, R.A. (Eds.). (2019). *The Cambridge Handbook of Stakeholder Theory*, Cambridge University Press. https://doi.org/10.1017/9781108123495.

Heravi, A., Coffey, V., and Trigunarsyah, B. (2015). Evaluating the level of stakeholder involvement during the project planning processes of building projects, *International Journal of Project Management*, **33**(5): 985–997. https://doi.org/10.1016/j.ijproman.2014.12.007.

Hillman, A., and Keim, G.D. (2001). Shareholder value, stakeholder management and social issues: What's the bottom line? *Strategic Management Journal*, **22**(2): 125–139. https://doi.org/10.2307/3094310.

Jones, C., and Lichtenstein, B.B. (2008). Temporary inter-organizational projects. pp. 231–255. *In*: Cropper, S., C. Huxham, M. Ebers, and P.S. Ring (eds.). *The Oxford Handbook of Inter-Organizational Relations*. Oxford University Press. https://doi. org/10.1093/oxfordhb/9780199282944.003.0009.

Jones, T., and Wicks, A. (1999). Convergent stakeholder theory, *The Academy of Management Review*, **24**(2): 206–221.

Kaler, J. (2003). Differentiating stakeholder theories, *Journal of Business Ethics*, **46**(1): 1–83. https://link.springer.com/content/pdf/10.1023%2FA%3A1024794710899.pdf.

Kaptein, M., and Van Tulder, R. (2003). Toward effective stakeholder dialogue, *Business and Society Review*, **108**(2): 203–224.

Kietzmann, J.H., Hermkens, K., McCarthy, I.P., and Silvestre, B.S. (2011). Social media? Get serious! Understanding the functional building blocks of social media, *Business Horizons*, **54**(3): 241–251. https://doi.org/10.1016/j.bushor.2011.01.005.

Klein, P.G., Mahoney, J.T., McGahan, A.M., and Pitelis, C.N. (2019). Organizational governance adaptation: Who is in, who is out, and who gets what, *Academy of Management Review*, **44**(1): 6–27. https://doi.org/10.5465/amr.2014.0459.

Kujala, J., and Sachs, S. (2019). The practice of stakeholder engagement. pp. 227–241. *In:* Harrison, J.S., Barney, J.B., Freeman, R.E. and Phillips, R.A. (Eds.). *The Cambridge Handbook of Stakeholder Theory*, Cambridge University Press.

Lee, K., Oh, W.Y., and Kim, N. (2013). Social media for socially responsible firms: Analysis of fortune 500's twitter profiles and their CSR/CSIR ratings, *Journal of Business Ethics*, **118**(4): 791–806. https://doi.org/10.1007/s10551-013-1961-2.

Lehtimaki, H., and Kujala, J. (2017). Framing dynamically changing firm – stakeholder relationships in an international dispute over a foreign investment: A discursive analysis approach, *Business and Society*, **56**(3): 487–523. https://doi.org/10.1177/0007650315570611.

Lehtinen, J., Aaltonen, K., and Rajala, R. (2019). Stakeholder management in complex product systems: Practices and rationales for engagement and disengagement, *Industrial Marketing Management*, **79**: 58–70. https://doi.org/10.1016/j.indmarman.2018.08.011.

Lehtinen, J., and Aaltonen, K. (2020). Organizing external stakeholder engagement in inter-organizational projects: Opening the black box, *International Journal of Project Management*, **38**(2): 85–98. https://doi.org/10.1016/j.ijproman.2019.12.001.

Lehtinen, J. (2021). *External Stakeholder Engagement in Complex Projects*, Aalto University, Finland.

Liu, B., Li, Y., Xue, B., Li, Q., Zou, P.X.W., and Li, L. (2018). Why do individuals engage in collective actions against major construction projects? An empirical analysis based on Chinese data, *International Journal of Project Management*, **36**(4): 612–626. https://doi.org/10.1016/j.ijproman.2018.02.004.

Lobo, S., and Abid, A.F. (2020). The role of social media in intrastakeholder strategies to influence decision making in a UK infrastructure megaproject: Crossrail 2, *Project Management Journal*, **51**(1): 96–119. https://doi.org/10.1177/8756972819864456.

Lovejoy, K., and Saxton, G.D. (2012). Information, community, and action: How nonprofit organizations use social media, *Journal of Computer-mediated Communication*, **17**(3): 337–353. https://doi.org/10.1111/j.1083-6101.2012.01576.x.

Lundin, R.A., and Söderholm, A. (1995). A theory of temporary organization, *Scandinavian Journal of Management*, **11**(4): 437–455.

Markovic, S., and Bagherzadeh, M. (2018). How does breadth of external stakeholder co-creation influence innovation performance? Analysing the mediating roles of knowledge sharing and product innovation, *Journal of Business Research*, **88**(March): 173–186. https://doi.org/10.1016/j.jbusres.2018.03.028.

Martini, A., Massa, S., and Testa, S. (2014). Customer co-creation projects and social media: The case of Barilla of Italy, *Business Horizons*, **57**(3): 425–434. https://doi.org/10.1016/j.bushor.2014.02.003.

Maylor, H., and Turner, N. (2017). Understand, reduce, respond: Project complexity management theory and practice, *International Journal of Operations & Production Management*, **37**(8): 1076–1093. https://doi.org/10.1108/IJOPM-05-2016-0263.

Mitchell, R., Agle, B., and Wood, D. (1997). Toward a theory of stakeholder identification and salience: Defining the principle of who and what really counts, *Academy of Management Review*, **22**(4): 853–886.

Mok, K.Y., Shen, G.Q., and Yang, J. (2015). Stakeholder management studies in mega construction projects: A review and future directions, *International Journal of Project Management*, **33**(2): 446–457. https://doi.org/10.1016/j.ijproman.2014.08.007.

Monteiro de Carvalho, M. (2013). An investigation of the role of communication in IT projects, *International Journal of Operations & Production Management*, **34**(1): 36–64. https://doi.org/10.1108/IJOPM-11-2011-0439.

Nguyen, T.H.D., Chileshe, N., Rameezdeen, R., and Wood, A. (2019). External stakeholder strategic actions in projects: A multi-case study, *International Journal of Project Management*, **37**(1): 176–191. https://doi.org/10.1016/j.ijproman.2018.12.001.

Ninan, J., Clegg, S., and Mahalingam, A. (2019). Branding and governmentality for infrastructure megaprojects: The role of social media, *International Journal of Project Management*, **37**(1): 59–72. https://doi.org/10.1016/j.ijproman.2018.10.005.

Ninan, J., Mahalingam, A., Clegg, S., and Sankaran, S. (2020). ICT for external stakeholder management: Socio-materiality from a power perspective, *Construction Management and Economics*, **38**(9): 840–855. https://doi.org/10.1080/01446193.2020.1755047.

Noland, J., and Phillips, R. (2010). Stakeholder engagement, discourse ethics and strategic management, *International Journal of Management Reviews*, **12**(1): 39–49. https://doi.org/10.1111/j.1468-2370.2009.00279.x.

Passetti, E., Bianchi, L., Battaglia, M., and Frey, M. (2019). When democratic principles are not enough: Tensions and temporalities of dialogic stakeholder engagement, *Journal of Business Ethics*, **155**(1): 173–190. https://doi.org/10.1007/s10551-017-3500-z.

Phillips, R., Freeman, R.E., and Wicks, A.C. (2003). What stakeholder theory is not, *Business Ethics Quarterly*, **4**: 479–502. https://doi.org/10.2307/3857968.

Rasche, A., and Esser, D.E. (2006). From stakeholder management to stakeholder accountability, *Journal of Business Ethics*, **65**(3): 251–267. https://doi.org/10.1007/s10551-005-5355-y.

Reynolds, M., and Yuthas, K. (2008). Moral discourse and corporate social responsibility reporting, *Journal of Business Ethics*, **78**(1-2): 47–64. https://doi.org/10.1007/s10551-006-9316-x.

Savage, G.T., Nix, T.W., Whitehead, C.J., and Blair, J.D. (1991). Strategies for assessing and managing organizational stakeholders, *Academy of Management Perspectives*, **5**(2): 61–75.

Shymko, Y., and Roulet, T.J. (2017). When does medici hurt da vinci? Mitigating the signaling effect of extraneous stakeholder relationships in the field of cultural production, *Academy of Management Journal*, **60**(4): 1307–1338. https://doi.org/10.5465/amj.2015.0464.

Spence, L.J., and Rinaldi, L. (2014). Governmentality in accounting and accountability: A case study of embedding sustainability in a supply chain, *Accounting, Organizations and Society*, **39**(6): 433–452. https://doi.org/10.1016/j.aos.2012.03.003.

Steurer, R. (2006). Mapping stakeholder theory anew: From the 'Stakeholder Theory of the Firm' to three perspectives on business-society relations, *Business Strategy and the Environment*, **15**(1): 55–69. https://doi.org/10.1002/bse.467.

Turkulainen, V., Aaltonen, K., and Lohikoski, P. (2015). Managing project stakeholder communication: The qstock festival case, *Project Management Journal*, **46**(6): 74–91. https://doi.org/10.1002/pmj.21547.

Unterhitzenberger, C., Wilson, H., James Bryde, D., Rost, M., and Joby, R. (2020). The stakeholder challenge: Dealing with challenging situations involving stakeholders, *Production Planning & Control*, 1–16. https://doi.org/10.1080/09537287.2020.1776907.

Viglia, G., Pera, R., and Bigné, E. (2018). The determinants of stakeholder engagement in digital platforms, *Journal of Business Research*, **89**: 404–410. https://doi.org/10.1016/j.jbusres.2017.12.029.

Williams, N.L., Ferdinand, N., and Pasian, B. (2015). Online stakeholder interactions in the early stage of a megaproject, *Project Management Journal*, **46**(6): 92–110. https://doi.org/10.1002/pmj.21548.

Winch, G.M. (2004). Managing project stakeholders. pp. 321–339. *In:* Morris, P.W.G. and Pinto, J.K. (Eds.). *The Wiley Guide to Managing Projects*, https://doi.org/10.1002/9780470172391.ch14.

Winch, G. (2017). Megaproject stakeholder management. pp. 339–361. *In*: Flyvbjerg, B. (ed.). *The Oxford Handbook of Mega-project Management*. Oxford University Press.

Zhang, Y., Sun, J., Yang, Z., and Wang, Y. (2018). Mobile social media in inter-organizational projects: Aligning tool, task and team for virtual collaboration effectiveness, *International Journal of Project Management*, **36**(8): 1096–1108. https://doi.org/10.1016/j.ijproman.2018.09.003.

10

Developing a Social Media Unit for Project Management Students

Arash Daemi, Ritesh Chugh* and *Muralitheran V Kanagarajoo*

10.1 Introduction

Since its inception, social media has evolved in leaps and bounds globally. Social media platforms initially used for socialising by individuals are now used for, but are not limited to, communication, collaboration, seeking opinions, brand monitoring, and marketing by organizations. Social media has become a popular tool for individuals and businesses alike, offering a wide range of useful and free features. As the growth and popularity of social media inspire the introduction of newer platforms and features, the emergence of new options can result in the broader usage of social media in different domains. Project management is one of the domains where businesses could benefit from the use of social media. The usage of social media for requirement management, communication management, knowledge management, policymaking, and collaboration in projects is documented in the literature. For example, social media features can offer flexibility and transparency in the requirement management of stakeholders (Sundaramoorthy and Bharathi, 2016). Social media's broad communication channels can facilitate easier access to internal and external stakeholders. The instant messaging feature enables project managers to communicate swiftly on project issues (Hysa and Spalek, 2019). The

Central Queensland University, Australia.
* Corresponding author: a.daemi@cqu.edu.au

embedded potential to share files can boost efficiency and effectiveness of communication and the decision-making process (Kanagarajoo et al., 2019; Orta-Castañon et al., 2018). Similar features of social media platforms can be employed to manage a project knowledge hub, such as establishing knowledge repositories for cross-functional teams (Sarka and Ipsen, 2017) or disseminating and retrieving lessons learned.

Despite various benefits and applications, the literature suggests that social media's adoption in project management has been lacklustre (Hysa and Spalek, 2019; Kanagarajoo et al., 2019). The lack of competency and confidence for appropriate and secure use of social media in project management inhibits its use (Rimkunienė and Zinkeviciute, 2014). Although social media offers a variety of features, inappropriate use can prove to be problematic for organizations. Often the problems in social media usage stem from a lack of skills for effective use in organizational contexts. Hence, raising awareness through training and education is suggested to overcome the competency gap. Furthermore, training time, cost, and resources are also identified as reasons for hindering social media adoption in project management. By integrating social media content in the project management curriculum, we aim to address the above issues holistically.

To the best of our knowledge, no curriculum has been developed to train and prepare project management students for informed and competent use of social media. As a standard practice, the project management curriculum includes describing and applying project management tools and techniques in diverse scenarios. However, students are not taught how to benefit from different social media features, especially in a project management context. Additionally, the project management curriculum in higher education needs to adapt to millennial learner (Gen Y) characteristics. With high technology usage patterns, these cohorts frequently seek and acquire information and content from diverse sources. They are social media savvy and use social communication tools. Such characteristics need to be considered and embedded in the project management curriculum too. There are examples of similar considerations in marketing and medical curriculum (Bahner et al., 2012; Crittenden and Crittenden, 2015; Faulds and Mangold, 2014; Mesko, Győrffy and Kollár, 2015) where students are taught to use social media features in different scenarios, along with compliance with guidelines. Project management students also need to be taught how to use social media. Furthermore, project management's reliance on communication and collaboration, cross-boundary social media features, frequent technology updates, and the learning profile of millennials substantiate a revamped project management curriculum that includes a social media unit in educational settings.

However, there appear to be no units that specifically cater to teaching social media use to project management students. Since no prior attempt has been made to address this gap in higher education's current project management curriculum, this chapter aims to propose a unit structure for teaching social media use to project management students.

10.2 Literature Review

In recent years, there has been a growing trend in the use of social media for business operations. Some companies utilize publicly available free social media platforms, whereas others invest in enterprise social media systems (Oostervink et al., 2016) that provide additional features catering to organizational use, better privacy and security of information. There are many reasons for the immense popularity of social media use in business environments. While not replacing the ideal in-person mode for communication, social media can facilitate a connection between groups of various sizes from multiple geographical regions and time zones (Kanagarajoo et al., 2019). This is particularly beneficial given today's global economy and the likelihood of engaging in international business. On the other hand, social media makes file sharing easy, particularly for images, videos, and large volume content (Orta-Castañon et al., 2018). We can add various possibilities to these benefits, from video calls to text and multimedia messages and the simplified ease of access to data within a platform. It also provides the opportunity to easily arrange and organize virtual groups, whether internal or external stakeholders, for general communication or specific objectives. Such functionalities facilitate fast and easy cross-boundary communication, enhancing efficiency, team collaboration, and innovation (Hysa and Spalek, 2019; Orenga-Roglá and Chalmeta, 2017; Pivec and Maček, 2019; Wang and Medaglia, 2017). These features connote social media's potential benefits in managing business operations, including projects and project management in general.

10.2.1 Social Media Competency and Education Providers

While numerous benefits of using social media in business environments are documented, organizations often fail to benefit from the full prospect of benefits due to diverse human factors. For example, varying levels of digital literacy, employee reservations in trying new digital platforms or technology (Young and Hinesly, 2014) entail incompetency and scepticism in the appropriate and secure use of social media (Rimkuniené and Zinkeviciute, 2014). In addition to the above barriers, restricted time, cost, and resources for upskilling human resource competencies refrain organizations to conquer these barriers internally. These barriers signify education providers' role in preparing the future workforce and equipping them with social media usage and application skills.

In a world heavily influenced by information and communication technology (ICT), the effective integration of ICT in pedagogy has proved to be vital in training learners as well as the general interest from industry in almost every domain. Consequently, the knowledge and competency required to integrate social media have now become imperative in the academic world (Megele, 2015). A large number of education providers have recently focused on developing and delivering courses on the subject of social media use for educational purposes (Romero-Hall and Li, 2020). These courses intend to educate and upskill learners on the prospects of using social media as a tool for collaborative learning. Although this approach integrates social media use in educational programmes, it only embeds social media into the education context for teaching and assessment purposes. Despite the popularity of social media and the flourishing concept of using social media as an educative tool, only limited studies have focused on the content of courses teaching social media use in a given profession (Romero-Hall and Li, 2020). In this study, we aim to introduce a content structure for teaching project management students on using social media for project management activities. However, it is essential to initially review the considerations influencing the unit design, which would subsequently frame the unit content.

10.2.2 Considerations Influencing the Unit Design

A general expectation from any higher education provider is to provide an outstanding educational experience to learners. An outstanding experience, in this case, is defined by a personalized teaching experience that caters to various student learning approaches, a supportive environment, and a learning journey that would successfully embed in student life and result in high success rates (Cooper and Krishnan, 2020). The fulfilment of such experiences depends on the course and unit designers, curriculum developers, content developers and education experts, and teaching professionals. These experts identify and use the most effective, engaging, and personalisable teaching and curriculum development approaches and learning models. Hence, first, we review recent curriculum development approaches and contrast some learning models.

The fundamental changes in curricula development result from extensive research into student learning and development patterns and close observation of the correlation between the teaching methodology, course content, and student development (Entwistle and Entwistle, 1970). This approach advocates rethinking on the design of the course content in both unit and course levels to bring in the broader context of the course content and real-life application of the knowledge gained. In this transition, higher education providers have moved from the broken discourse between an isolated course designer and a disconnected team of teachers or lecturers to acknowledging and embracing a cumulative and progressive process of learning; consequently, engaging students in problem-based, project-based, case-based, and scenario-based learning activities inspired by active learning concepts (Cooper and Krishnan, 2020). In the didactic model, knowledge is seen as an external component to the learner. In this model, the owner of the knowledge (teacher) transfers the abstract knowledge to the learner, who in turn receives it via cognitive approaches. On the other hand, the social learning model places the best opportunities for learning and engagement in the social exchange of knowledge, often through action-based learning. Social learning theory is entirely different from the didactic learning model in shifting the learning process from a passive teacher-dominated context to a highly engaging student-centred one, where the learner actively takes part in the learning process by guided action learning and practice (Megele, 2015). Action-based learning is an enquiry-driven approach in which the learner engages in a problem and where learning occurs through exploring and resolving problems (Revans, 1982). Previous research notes that learning that occurred through problem-based approaches leaves the learner in a better place in tackling real-world problems in comparison to students subject to traditional education (Megele, 2015).

In addition to different learning models, learners' attributes are another concept to consider in the unit design. In the current education environment, we face a learner generation (also referred to as the Net generation) with distinct characteristics, such as digital literacy, immediacy, connectivity, interactivity, collaboration, experiential and engaged learning (Skiba and Barton, 2006). The Net generation learners' characteristics also include lack of oral and written communication skills, incompetency in critical analysis where they need to evaluate the accuracy of information and consequently deficiency in problem-solving skills (Faulds and Mangold, 2014). Such a cohort of learners presents a unique set of challenges to the concept of learning design and curriculum development. Faulds and Mangold (2014) implemented strategies to overcome the challenges these learners present in an educational environment by using highly-structured learning activities, providing comprehensive immediate feedback, adopting experiential learning, incorporating visual elements, thereby adding the element of excitement and creativity. A distinct approach they used in tackling these challenges is creating and incorporating learning communities in classroom activities (Faulds and Mangold, 2014).

The experiential or reflection model of course design corresponds with the action-based model and the reviewed characteristics of learners. The experiential or reflection model is an approach where the course is designed to stay as closely in touch with the real-life situations relevant to the subject matter as possible. In this design approach, adult learners attempt authentic assessments to solve problems and learn from experience (O'Neill, 2015). The experiential or reflection model is a student-centered process that aims to keep the learning process as a challenge and the learners constantly active (Faulds and Mangold, 2014). Similarly, outcome-based learning has become the most dominant trend of curriculum design in higher education for some time now (Cooper and Krishnan, 2020). Referred to as a paradigm shift, the trend aims to link the theoretical knowledge and application of skills in context for the learners and graduates. This, in turn, provides them with job-ready skills and upskills learners to perform and succeed in the real-world context of the application of their education than being limited by theory (Cooper and Krishnan, 2020). The reviewed design models highlight the importance of integrating academic knowledge and real-life experience.

Integrating theory and practice as a mandatory characteristic of an effective curriculum design has forced higher education providers to implement drastic changes in their curriculum design. Having to withstand and perform to the standards of a rapidly changing world and its political and economic trends and requirements for a demand-driven skilled workforce, the curriculum has been changed to allow graduates to develop skills and knowledge vital for effective integration in the current and future workplace (Cooper and Krishnan, 2020). Integrative learning can be defined as associating the theory, knowledge, and skills with practice (O'Neill, 2015). Through integrative learning, students explore how different people involved in complex situations view, perceive, and understand these situations differently (Newell, 2010). This approach encourages students to develop connections between the learned concepts and new or complex issues. In summary, an integrated curriculum would allow the learner to connect and apply his/her skills and knowledge in the subject matter context, to identify or develop and apply variously suitable and pragmatic approaches and ideas or a combination of them to a real-life context (O'Neill, 2015). To achieve such abilities, a learner would require consolidation of the gained knowledge and skills, which is the factor that links integrity in curriculum design to continuity.

Moreover, integrative learning correlates to continuity. Unlike modularisation, criticized for its isolation from integrative learning, continuity in curriculum design allows the learner to consolidate the knowledge and provides opportunities for integrated learning by applying knowledge through various activities to contextual scenarios (O'Neill, 2015). Continuity in the curriculum can benefit the learner not only in the prospect of deepening the knowledge on the subject matter, but also further developing the necessary associated skills, such as team working and problem-solving. Assessment tools of a continuous curriculum should equally carry the same notion. They would ideally require the learner to exercise a higher level of subject matter knowledge as well as more advanced cognitive competencies, such as judging, comparing, debating, and self and peer assessment in a more autonomous way (O'Neill, 2015).

10.3 Backward Design Process Model

Reflecting on the literature and our own experience as project management educators, we believe that appropriate diagnosis of graduate attributes can be the cornerstone of educational programmes design. This approach enables designers to consider the required knowledge and skills in their design so that graduates can effectively function in the corresponding profession. This also aligns with Wiggins and McTighe (1998)' focus on students' needs in their Backward Design Process, comprising of three stages: 1. identify desired results, 2. determine acceptable evidence, and 3. plan learning experiences and instructions. This process has been used for the design of learning programmes (Najdanovic-Visak, 2017; Shah et al., 2018) and is adapted for this study as shown in Fig. 10.1 and as explained below.

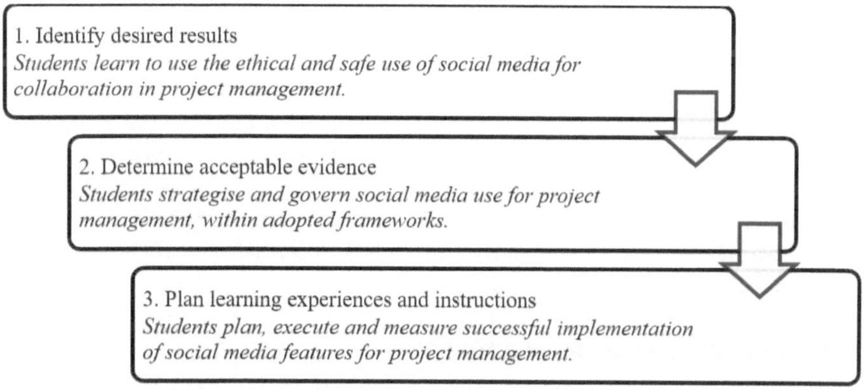

Fig. 10.1: The Backward Design Process applied to a project management under/postgraduate unit.
(Alt Text: The backward design process coving three steps—identify desired results, determine acceptable evidence, and plan learning experiences and instructions)

10.3.1 Identify Desired Results

In this stage, educational designers need to establish the desired goal and the extent to which the associated topics are explored, resulting in enduring understanding in learners (Wiggins and McTighe, 1998). As discussed earlier, this study aims to address the lack of competency and

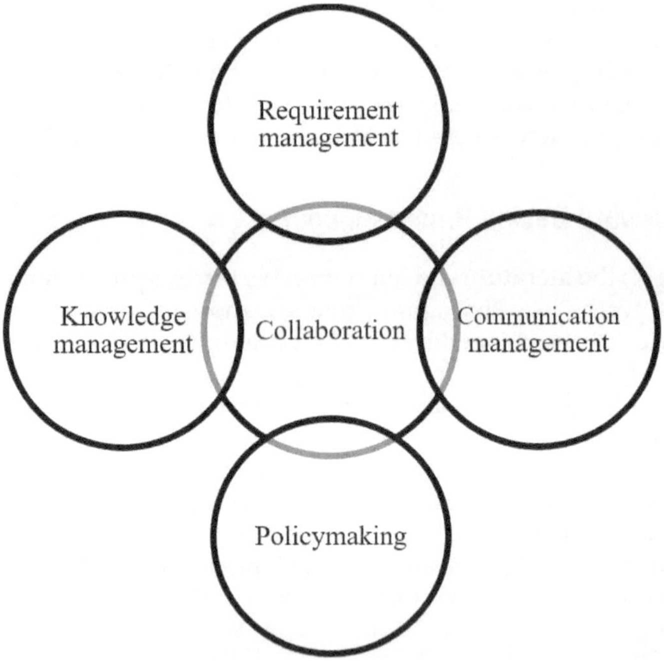

Fig. 10.2: Project management areas where social media benefits.
(Alt Text: Social media benefits to project management such as collaboration, requirements management, communication management, policy making, and knowledge management)

confidence in the appropriate and secure use of social media in project management. Moreover, as Fig. 10.2 illustrates, collaboration overlaps with other areas in which social media use in projects is recorded in the literature (Daemi et al., 2020). Thus, ethical and safe use of social media for collaboration in project management focuses on enduring understanding for learners.

10.3.2 Determine Acceptable Evidence

In the second stage, designers need to determine the evidence that proves learners understand and are proficient in the previously established desired results (Wiggins and McTighe, 1998). Strategizing and governing social media use for project management in compliance with adopted frameworks can provide evidence of the ethical and safe use of social media for collaboration in project management. Adopted frameworks can vary from international to regional standards at both organizational and individual levels, including different policies and procedures. The compliance requirement foresees the ethical and safe application of social media use in the provided evidence. Subsequently, strategizing and governing social media use for project management requires learners to review the social media use objectives and collaborate effectively with corresponding stakeholders in project environment. For such evidence, performance tasks and projects are good options in the choice of assessment (Wiggins and McTighe, 1998).

10.3.3 Plan Learning Experiences and Instructions

Keeping in mind the previously established desired results and required evidence, in this stage, designers plan supporting materials and instructions, enabling the intended knowledge and skills in learners (Wiggins and McTighe, 1998). Based on the previous stages, learners need the knowledge and skills for planning, executing, and measuring successful implementation of social media features in project management. Such knowledge and skills can aid learners to effectively demonstrate their understanding and proficiency in social media use for project management. Thus, in the following sections, we propose instructions and unit content that can aid learners to acquire the required knowledge and skills for ethical and safe use of social media for collaboration in project management.

10.4 A Unit Design for Integrating Social Media in Project Management Practices

While the mentioned gaps in the project management curriculum and industry are noticeable, it can be argued that teaching social media applications in project management to project management students can be helpful. As earlier discussions highlighted the benefits of social media use in project management, we stress the importance of reconnecting the academically focused project management studies with the reality of the industry. Equivalently, we underline the significance of closing the gap between the learner skillset and the necessities for success as current project management professionals. Hence, a suitable unit design is needed to teach the application of social media usage skills to project management students. Based on the Backward Design Process discussion, the following topics and those in the discussion section are suggested to be covered in the 'social media for project management' unit.

10.4.1 Social Media

As this unit discusses different concepts around the application and governance of social media in project management, reviewing social media definitions and characteristics would be beneficial for students to inform future decisions. Carr and Hayes (2015) contrast different social media definitions while suggesting their own as the 'internet-based, disentrained, and persistent channels of mass personal communication facilitating perceptions of interactions among users, deriving value primarily from user-generated content' (p.49). Later, Chugh and Ruhi (2019) suggested their functional definition of social media as 'a web-based technological platform for online communication that enables users to connect and interact, share ideas and information, build and engage social networks, and develop online communities' (p.1).

Besides the definition, social media functionality is of importance to guide the strategies to perform and monitor social media activities (Kietzmann et al., 2011). For instance, Kietzmann et al. (2011) discuss social media functionality and implications of the functionality through the seven building blocks of their honeycomb model. The blocks (identity, conversations, sharing, presence, relationships, reputation, and groups) represent vital aspects of user engagement on social media, alongside the examination of user experience. Such knowledge can reinforce students' insight into appropriate and effective use of social media.

10.4.2 Develop, Execute and Monitor a Social Media Adoption Strategy for Project Management

A social media adoption strategy is the key contributor to the effective use of social media (Daemi et al., 2020; Duane and O'Reilly, 2017; Hysa and Spalek, 2019; Kanagarajoo et al., 2019). Furthermore, social media strategy needs alignment with the organizational objectives for social media use (Daemi et al., 2020). Therefore, in this unit, developing a social media strategy is crucial for influencing the positive use of social media in project management.

We first recommend a review and analysis of setting social media objectives in organizations. This will inform the next step, which determines social media requirements. Once social media use objectives and requirements are established, they shape the social media application framework in conjunction with appropriate policies and procedures. In the following steps, a plan for social media use should be developed, executed, and monitored within the established framework. Periodic evaluation of the social media effectiveness enables organizations to adjust each of the previous steps to yield social media adoption objectives.

10.4.3 Identify and Review Social Media Platforms for Organizational Use

Although social media refers to various Web 2.0 platforms, each social media platform is associated with a certain number of characteristics that may or may not be suitable for a certain type of function and objectives. For instance, Kietzmann et al. (2011) have reviewed some social media platforms and determined the type of functionality each platform provides. In developing a social media adoption strategy, similar steps should be taken to identify the most suitable platforms for social media adoption objectives. This also includes an investigation on enterprise social media platforms.

The initial step should equip students with the knowledge of sourcing information on social media platforms. This may include investigative knowledge and techniques for sourcing the required information. Utilizing similar tools and techniques, each platform's functions and features are learned. Before selecting suitable platforms, it is recommended to review industry-specific requirements and preferences. The captured information should enable identifying suitable social media platforms for meeting the agreed objectives.

10.4.4 Use Social Media Tools for Collaboration

Before using social media tools, it would be worthwhile to learn about the target audience. While for some people using social media may translate to using new tools and acquiring new skills and mindset (Cardon and Marshall, 2015), it might be second nature for others. Earlier, we outlined millennial characteristics and the importance of considering these attributes during development. It is also worth noting that millennials are known to be the least engaged workforce, and social media use helps improve their engagement level (Naim and Lenka, 2017).

Once preliminary factors are considered, we recommend describing and contrasting different social media tools and applications. Such knowledge enables students to choose their tools effectively when a scenario is presented to them. Students will learn how to set up and use popular social media tools and applications for organizational use in the corresponding step.

10.4.5 Manage Safe Use of Social Media

Concern over information security is a key barrier of social media adoption in organizations (Hysa and Spalek, 2019; Kanagarajoo et al., 2019; Picazo-Vela et al., 2012; Rimkunienė and Zinkeviciute, 2014). Furthermore, security breaches can negatively impact organizational economy (Zhang and Gupta, 2018). Scholars have reviewed security risks of social media use (Gritzalis et al., 2014; Patel and Jasani, 2010; Picazo-Vela et al., 2012; Rose, 2011; Zhang and Gupta, 2018) and appropriate mitigation techniques, including identifying the education of workforce (He, 2012). Thus, we recommend a section in which students learn about social media risks and undertake appropriate measures for prevention or mitigation.

For an effective risk management practice, risk management processes are reviewed first. Then potential risks associated with social media use are reviewed, and appropriate risk management strategies for each risk are proposed. Finally, students learn how to formulate, execute, monitor, and evaluate risk management plans.

10.4.6 Maintain Ethical and Professional Standards when Using Social Media Platforms

There are various perspectives on social media ethics, for instance, in project management, social media can be used for stakeholders' engagement (Ninan et al., 2019; Ninan et al., 2020; Zhou et al., 2021). Bowen (2013) highlights the importance of upholding accuracy, honesty, and full disclosure when disseminating messages. From another perspective, maintaining personal and professional boundaries are underlined and answered by Kimball and Kim (2013). Scholars suggest educating and creating an increased awareness to better understand and avoid unethical social media practices (Fox and Bird, 2017; Osborne et al., 2015). Thus, we recommend a section to address the above concerns.

As a starting point, students learn to research and analyze the general impacts of social media. Then, governing ethical and professional codes of conduct for social media use and online presence are reviewed at industry and organization levels. Such knowledge will inform necessary compliance decisions and actions. Furthermore, the fundamentals of protecting individuals' privacy and organizations' confidential information are covered. Students then learn and practice to engage with internal and external stakeholders.

10.4.7 Analyze Performance of Social Media Tools

Like other initiatives for managing projects, social media tools also require assessment. The evaluation must align with the objectives of social media use in project management to appropriately inform required adjustments. Therefore, we recommend teaching students some methods of social media performance analysis. For example, user engagement on social media posts is determined by analyzing likes, comments and shares (Chugh et al., 2019). More specifically, this includes measuring the time spent on content (video), the number of visits, and the number of likes or clicks on content (Saura et al., 2021).

At first, students learn about the tools used for performance analysis and the way they function. Then, data collection methods and tools are described and contrasted for increased insight into the purpose and capacity of each. Practising data collection, analysis, and performance assessment can be arranged next to reinforce students' cognitive learning.

In addition to the above-discussed content, specifically designed workshop activities, practical projects and assessments are needed to fulfil integrative learning and continuity.

10.5 Discussion

Rethinking the design of the project management course content at both unit and course level due to globalization and human factors constituting current projects is vital. Earlier in this study, we suggested social media-related topics to be covered in the unit design. In addition to those topics, we recommend providing an overview of the following topics as they influence how social media would be used in a project.

10.5.1 Project Management Practices

This unit design aims to educate project management students on utilizing and governing social media in project management. The Project Management Body of Knowledge (PMBOK) is the chosen project management framework for this unit's design. PMI (2017) describes PMBOK 'as a foundation upon which organizations can build methodologies, policies, procedures, rules, tools and techniques, and life cycle phases needed to practice project management' (p. 2). While this unit focuses on integrating social media tools into project management practices that are aligned with the PMBOK, students must be made aware of the influence of diverse project management practices on social media use. There are examples where scholars have contrasted different project management practices, such as PMBOK, ICB, PRINCE2, AGILE, APMBOK, P2M, ISO 10006, and ISO 21500 (Abdullah et al., 2021; Al-Maghraby, 2010; Drob and Zichil, 2013). Such studies, for instance, can inform students how communication objectives and governance may vary in different project management practices and consequently the influence they may have on social media use.

10.5.2 Project Management Lifecycle Models

A project life cycle depicts a project's stages to produce and approve its deliverables (Assudani and Kloppenborg, 2010). Concept, definition, execution and closeout are the broad terms used for project phases based on a general consensus (Archibald, 2013). As project management students learn to use social media, they need the understanding of the influence of project management life cycle on different project aspects (such as managing stakeholders) and subsequently success (Assudani and Kloppenborg, 2010; Bryde, 2003; Olsson and Magnussen, 2007; Pervoukhin et al., 2020).

Additionally, different project management life cycle models cater for different project demands (Archibald, 2013), and hence they require different governance models. For instance, in Waterfall and Agile life cycle models, the approach towards planning and the planning horizon differs to accommodate the factors influencing the predictability of projects. Furthermore, life cycle models differ in stakeholders' engagement, communication strategies and techniques, and factors contributing to organizational collaboration. Hence, as preliminary knowledge, this unit design recommends a brief overview of project management life cycle models and their influence on social media use.

10.5.3 Collaboration in Project Management

One of the major topics this unit focuses on is the application of social media to boost collaboration in projects. Collaboration can be translated into the stakeholders' interaction and collective actions towards a common goal, subsequently influencing the project performance and success (Bond-Barnard et al., 2018; Chiocchio et al., 2011). In current times when international and offshore projects bridge over geographical boundaries, the role of collaboration is imminent between multicultural and multilingual teams assembled from different time zones (Binder, 2016). The frequent emergence of such projects magnifies the importance of external and off-site collaboration (Qureshi et al., 2006).

However, before educating project management students on using and governing social media for improved project collaboration, we recommend a brief overview of the collaboration concept in project management. This will inform students about contributing and impeding factors to improved project collaboration as well as the theories influencing team and stakeholder dynamics. With such knowledge, students can choose appropriate social media features and define effective social media governance models for improved project collaboration.

10.5.4 The Unit Design

So far, we have proposed the topics to be covered in the social media for project management unit to prepare project management students for the appropriate and safe use of social media. Figure 10.3 illustrates the suggested order and design of the unit comprising of ten sections, starting with the project management practices and finishing with analysing the performance of social media tools and the efficacy of social media content.

The first three sections focus on how different project management approaches may influence the choice and use of social media. The rest of the sections gradually tend to build an understanding of social media

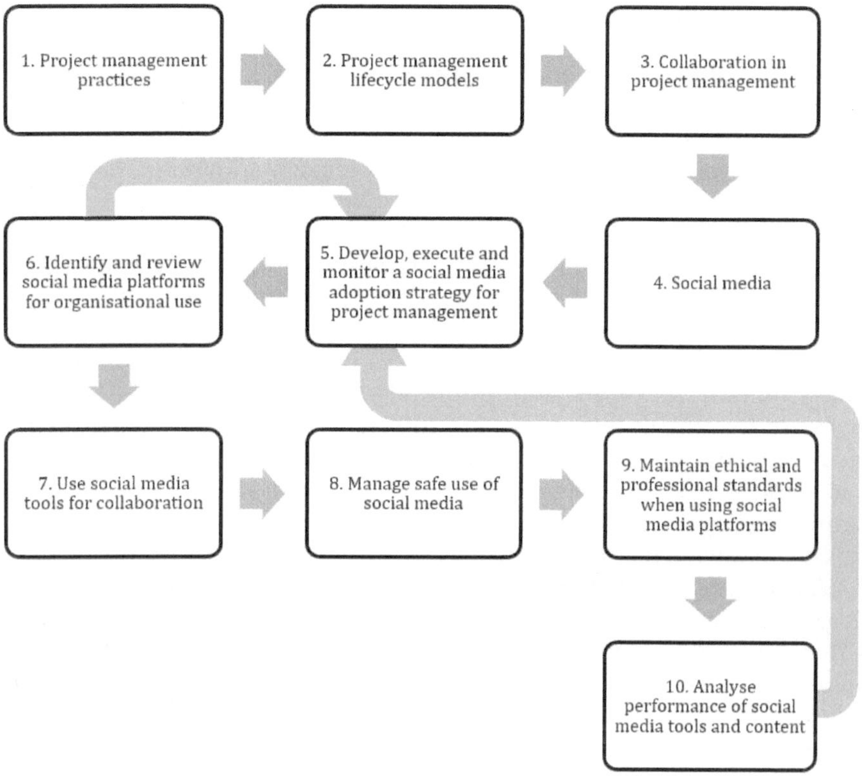

Fig. 10.3: The unit design.
(Alt Text: The ten steps to design a social media unit for project management shown as a flowchart)

and its application in the project management context. Through the last six sections, students will learn social media application and governance skills. In section five, students need to be reminded that for effective development, execution and monitoring of social media adoption in project management, they need inputs from sections six and ten. In other words, identification of appropriate social media platforms (section six) and analysis of social media tools performance (section ten) need to be considered in the development of social media adoption strategy (section five). Furthermore, analysis of social media tools performance (section ten) is one of the components of monitoring social media adoption strategy (section five) topic. Similarly, we stress that the outlined sections should be referred to in a holistic approach rather than separate and fragmented concepts.

10.6 Conclusion

Despite social media's benefits, organizations often fail to exploit its opportunities in project management activities due to human factors and in-house limitations in upskilling their personnel. This drawback suggests the role of education providers in preparing project management graduates to apply social media features in a project environment during their studies. However, our study underlined a gap in practice as there appeared to be no units that teach students about using social media in project management and for project management activities.

In this chapter, we call for a rethink in project management curriculum design to cater to current challenges imposed by international project boundaries to expanding virtual teams and millennial learning attributes shaping this era's learner profiles. To plug the gap, we used the Backward Design Process model to design a social media unit for project management students. Our unit design is the first attempt in this front, aiming to prepare project management students to use social media for project management and project activities.

As the first study that proposes integrating social media usage in the project management domain, it is limited in various ways. Firstly, it is theoretical in nature, and the proposed content requires testing and validation. In fact, it raises additional questions for future consideration. For example, should teaching social media applications be delivered in a stand-alone unit or across a project management course? Despite excluding this question from the scope of our study, we considered a stand-alone unit can better address the current gap in the industry, particularly as the content for the proposed unit is not expansive. However, the above question is worth being studied in future. Furthermore, the scope of our study did not include supplementary resources and assessment design. In addition, the implementation and evaluation of the proposed design can also be topics for future studies. After implementation, empirical studies can also test the efficacy of the unit's content. Finally, this chapter serves as a starting point for educators to consider adopting social media in their project management curriculum.

References

Abdullah, A.A., Abdul-Samad, Z., Abdul-Rahman, H., and Salleh, H. (2021). Project management methods, guides and standards: A critical overview, *Journal of Project Management Practice (JPMP)*, **1**(1): 35–51.

Al-Maghraby, R. (2010). Project management frameworks: Comparative analysis. Paper presented at the *Proceedings of the IPMA 2010 World Congress*, Istanbul, Turkey, pp. 1–4.

Archibald, R.D. (2013). A global system for categorising projects, *IPMA Project Perspectives*, **7**.

Assudani, R., and Kloppenborg, T.J. (2010). Managing stakeholders for project management success: An emergent model of stakeholders, *Journal of General Management*, **35**(3): 67–80.

Bahner, D.P., Adkins, E., Patel, N., Donley, C., Nagel, R., and Kman, N.E. (2012). How we use social media to supplement a novel curriculum in medical education, *Medical Teacher*, **34**(6): 439–444.

Binder, J. (2016). *Global Project Management: Communication, Collaboration and Management Across Borders*, Routledge.

Bond-Barnard, T.J., Fletcher, L., and Steyn, H. (2018). Linking trust and collaboration in project teams to project management success, *International Journal of Managing Projects in Business*, **11**(2): 432–457.

Bowen, S.A. (2013). Using classic social media cases to distill ethical guidelines for digital engagement, *Journal of Mass Media Ethics*, **28**(2): 119–133.

Bryde, D.J. (2003). Project management concepts, methods and application, *International Journal of Operations & Production Management*, **23**: 775–793.

Cardon, P.W., and Marshall, B. (2015). The hype and reality of social media use for work collaboration and team communication, *International Journal of Business Communication*, **52**(3): 273–293.

Carr, C.T., and Hayes, R.A. (2015). Social media: Defining, developing, and divining, *Atlantic Journal of Communication*, **23**(1): 46–65.

Chiocchio, F., Forgues, D., Paradis, D., and Iordanova, I. (2011). Teamwork in integrated design projects: Understanding the effects of trust, conflict, and collaboration on performance, *Project Management Journal*, **42**(6): 78–91.

Chugh, R., Patel, S.B., Patel, N., and Ruhi, U. (2019). Likes, comments and shares on social media: Exploring user engagement with a state tourism Facebook page, *International Journal of Web-based Communities*, **15**(2): 104–122.

Chugh, R., and Ruhi, U. (2019). Social media for tertiary education, *Encyclopedia of Education and Information Technologies*, 1–6, Springer Nature, Cham, Switzerland. doi.org/10.1007/978-3-319-60013-0_202-1.

Cooper, S.A., and Krishnan, S. (2020). *Effective Unit Design for Higher Education Courses: A Guide for Instructors*, Taylor & Francis.

Crittenden, V., and Crittenden, W. (2015). Digital and social media marketing in business education: Implications for the marketing curriculum, *Journal of Marketing Education*, **37**: 131–132.

Daemi, A., Chugh, R., and Kanagarajoo, M.V. (2020). Social media in project management: A systematic narrative literature review, *International Journal of Information Systems and Project Management*, **8**(4): 5–21.

Drob, C., and Zichil, V. (2013). Overview regarding the main guidelines, standards and methodologies used in project management, *Journal of Engineering Studies and Research*, **19**(3): 26–31.

Duane, A., and O'Reilly, P. (2017). A conceptual stages-of-growth model for managing a social media business profile, *Irish Journal of Management*, **36**(2): 78–98.

Entwistle, N.J., and Entwistle, D. (1970). The relationships between personality, study methods and academic performance, *British Journal of Educational Psychology*, **40**(2): 132–143.

Faulds, D.J., and Mangold, W.G. (2014). Developing a social media and marketing course, *Marketing Education Review*, **24**(2): 127–144.

Fox, A., and Bird, T. (2017). The challenge to professionals of using social media: Teachers in England negotiating personal-professional identities, *Education and Information Technologies*, **22**(2): 647–675.

Gritzalis, D., Kandias, M., Stavrou, V., and Mitrou, L. (2014). History of information: The case of privacy and security in social media, paper presented at the *Proc. of the History of Information Conference*.

He, W. (2012). A review of social media security risks and mitigation techniques, *Journal of Systems and Information Technology*, **14**(2): 171–180.

Hysa, B., and Spalek, S. (2019). Opportunities and threats presented by social media in project management, *Heliyon*, **5**(4): e01488. doi:10.1016/j.heliyon.2019.e01488.
Kanagarajoo, M.V., Fulford, R., and Standing, C. (2019). The contribution of social media to project management, *International Journal of Productivity and Performance Management*, **69**(4): 834–872. doi:10.1108/ijppm-09-2018-0316.
Kietzmann, J.H., Hermkens, K., McCarthy, I.P., and Silvestre, B.S. (2011). Social media? Get serious! Understanding the functional building blocks of social media, *Business Horizons*, **54**(3): 241–251.
Kimball, E., and Kim, J. (2013). Virtual boundaries: Ethical considerations for use of social media in social work, *Social Work*, **58**(2): 185–188.
Megele, C. (2015). eABLE: Embedding social media in academic curriculum as a learning and assessment strategy to enhance students learning and e-professionalism, *Innovations in Education and Teaching International*, **52**(4): 414–425.
Mesko, B., Győrffy, Z., and Kollár, J. (2015). Digital literacy in the medical curriculum: A course with social media tools and gamification, *JMIR Medical Education*, **1**(2): e4411.
Naim, M.F., and Lenka, U. (2017). The impact of social media and collaboration on Gen Y employees' engagement, *International Journal of Development Issues*, **16**(3): 289–299.
Najdanovic-Visak, V. (2017). Team-based learning for first year engineering students, *Education for Chemical Engineers*, **18**: 26–34.
Newell, W.H. (2010). Educating for a complex world: Integrative learning and interdisciplinary studies, *Liberal Education*, **96**(4): 6–11.
Ninan, J., Clegg, S., and Mahalingam, A. (2019). Branding and governmentality for infrastructure megaprojects: The role of social media, *International Journal of Project Management*, **37**(1): 59–72.
Ninan, J., Mahalingam, A., Clegg, S., and Sankaran, S. (2020). ICT for external stakeholder management: Sociomateriality from a power perspective, *Construction Management and Economics*, **38**(9): 840–855.
O'Neill, G. (2015). Curriculum design in higher education: Theory to practice, University College Dublin, *Teaching and Learning*.
Olsson, N.O., and Magnussen, O.M. (2007). Flexibility at different stages in the life cycle of projects: an empirical illustration of the 'freedom to manoeuvre', *Project Management Journal*, **38**(4): 25–32.
Oostervink, N., Agterberg, M., and Huysman, M. (2016). Knowledge sharing on enterprise social media: Practices to cope with institutional complexity, *Journal of Computer-Mediated Communication*, **21**(2): 156–176. doi:10.1111/jcc4.12153.
Orenga-Roglá, S., and Chalmeta, R. (2017). Methodology for the implementation of knowledge management systems 2.0, *Business & Information Systems Engineering*, **61**(2): 195–213. doi:10.1007/s12599-017-0513-1.
Orta-Castañon, P., Urbina-Coronado, P., Ahuett-Garza, H., Hernández-de-Menéndez, M., and Morales-Menendez, R. (2018). Social collaboration software for virtual teams: Case studies, *International Journal on Interactive Design and Manufacturing (IJIDeM)*, **12**(1): 15–24. doi:10.1007/s12008-017-0372-5.
Osborne, N., Connelly, L., Mesquita, A., and Peres, P. (2015). Managing your digital footprint: Possible implications for teaching and learning, paper presented at the *Proceedings of the 2nd European Conference on Social Media ECSM*, Porto, Portugal.
Patel, N., and Jasani, H. (2010). Social media security policies: Guidelines for organizations, *Issues in Information Systems*, **11**(1): 628–634.
Pervoukhin, D.V., Isaev, E.A., Rytikov, G.O., Filyugina, E.K., and Hayrapetyan, D.A. (2020). Theoretical comparative analysis of cascading, iterative, and hybrid approaches to IT project life cycle management, *Business Informatics*, **14**(1): 32–40.

Picazo-Vela, S., Gutiérrez-Martínez, I., and Luna-Reyes, L.F. (2012). Understanding risks, benefits, and strategic alternatives of social media applications in the public sector, *Government Information Quarterly*, **29**(4): 504–511.

Pivec, M., and Maček, A. (2019). Employment background influence on social media usage in the field of European project management and communication, *Journal of Business Research*, **94**: 280–289. doi:10.1016/j.jbusres.2018.03.021.

PMI. (2017). *A Guide to the Project Management Body of Knowledge (PMBOK® Guide)* (6th ed.): Newtown Square, PA: Project Management Institute.

Qureshi, S., Liu, M., and Vogel, D. (2006). The effects of electronic collaboration in distributed project management, *Group Decision and Negotiation*, **15**(1): 55–75.

Revans, R.W. (1982). What is action learning? *Journal of Management Development*, **1**(3): 64–75.

Rimkuniené, D., and Zinkeviciute, V. (2014). Social media in communication of temporary organizations: Role, needs, strategic perspective, *Journal of Business Economics and Management*, **15**(5): 899–914. doi:10.3846/16111699.2014.938360.

Romero-Hall, E., and Li, L. (2020). A syllabi analysis of Social Media for teaching and learning courses, *Journal of Teaching and Learning*, **14**(1): 13–28.

Rose, C. (2011). The security implications of ubiquitous social media, *International Journal of Management & Information Systems (IJMIS)*, **15**(1): 35–40.

Sarka, P., and Ipsen, C. (2017). Knowledge sharing via social media in software development: A systematic literature review, *Knowledge Management Research & Practice*, **15**(4): 594–609. doi: 10.1057/s41275-017-0075-5.

Saura, J.R., Palacios-Marqués, D., and Iturricha-Fernández, A. (2021). Ethical design in social media: Assessing the main performance measurements of user online behaviour modification, *Journal of Business Research*, **129**: 271–281.

Shah, V., Kumar, A., and Smart, K. (2018). Moving forward by looking backward: Embracing pedagogical principles to develop an innovative MSIS programme, *Journal of Information Systems Education*, **29**(3): 139–156.

Skiba, D.J., and Barton, A.J. (2006). Adapting your teaching to accommodate the net generation of learners, *Online Journal of Issues in Nursing*, **11**(2).

Sundaramoorthy, V., and Bharathi, B. (2016). Need for social media approach in software development, *Indian Journal of Science and Technology*, **9**(21): 1–6. doi: 10.17485/ijst/2016/v9i21/95271.

Wang, C., and Medaglia, R. (2017). Governments' social media use for external collaboration, *Transforming Government: People, Process and Policy*, **11**(4): 572–595. doi:10.1108/tg-02-2017-0009.

Wiggins, G., and McTighe, J. (1998). What is backward design? *Understanding by Design*, **1**: 7–19.

Young, A.M., and Hinesly, M.D. (2014). Social media use to enhance internal communication: Course design for business students, *Business and Professional Communication Quarterly*, **77**(4): 426–439.

Zhang, Z., and Gupta, B.B. (2018). Social media security and trustworthiness: Overview and new direction, *Future Generation Computer Systems*, **86**: 914–925.

Zhou, Z., Zhou, X., and Qian, L. (2021). Online public opinion analysis on infrastructure megaprojects: toward an analytical framework, *Journal of Management in Engineering*, **37**(1): 04020105.

11

Responsible Project Data Analytics

The Case of Social Media

Nigel Williams,[1,*] *Nicole Ferdinand*[2] and *Beverly Pasian*[3]

11.1 Project Social Media Data Analytics

Research and practice in the area of analytics are rapidly growing in prominence across businesses. Business analytics is defined as the use of emerging data sources and techniques to support evidence-based decision-making (Seddon and Currie, 2017). Therefore, project analytics can be defined as the use of project data sources to support decision-making in temporary organizations (projects, programmes, and portfolios) (Apm.org.uk). The volume of data available to project organizations has increased exponentially in the last ten years, and traditional analytics approaches cannot provide meaningful insights. These emerging data sources have distinct characteristics—volume (size of the dataset), velocity (rate of data collection and analysis), variety (heterogeneity of data sources), value (value of the processed data is low compared to its volume), and variability (time-sensitivity) (Seddon and Currie, 2017). In addition to internal sources, such as BIM, more significant organizational, social, and technological trends have increased the amount of data available to project organizations.

[1] University of Portsmouth Business School, United Kingdom.
[2] Oxford Brookes University, United Kingdom.
[3] Dortmund University of Applied Sciences, Germany.
* Corresponding author: nigel.williams@port.ac.uk

Projects now having a more comprehensive range of data, including temperature, pressure, Bluetooth signals, provide real-time insights into the context and location-based conditions. Social processes and business practices that help to provide the data and processing infrastructure sphere are often overlooked. New behavioral norms have been developed by individuals and organizations, in which specific social gratifications are derived from sharing data online (Liu and Bakici, 2019). Organizational business processes have changed to incorporate distributed sources of knowledge and practices, such as open data, open innovation, and open-source software (Guertler and Sick, 2021). Social sharing websites, such as Github, have been developed to enable the collaborative distribution, modification, and improvement of software code and programming practices, allowing rapid adoption of new data processing techniques (Cosentino et al., 2017). Public institutions have also embraced open data practices that encourage the sharing of data and research (Ruijer and Martinius, 2017).

While research studies use specific analytical techniques to generate academic insights, little effort has been made to examine how the project research community can incorporate these approaches to improve the research process. This is somewhat surprising as project work, in uncertain and temporally ambiguous environments in which sense-making is difficult, often results in suboptimal outcomes or project failure. A related issue is that the temporal nature of projects requires immediate, contextual insights since there is rarely a prolonged stable period for longitudinal analysis. Data may be valuable before particular milestones or decisions of little value have passed. Using this data to create project insights requires integrating data management processes, infrastructure, and human competencies. These insights can be generated via current use cases.

(1) Descriptive analytics use tools to present project data in a compelling format.
(2) Predictive analytics uses data to predict future project states or performance, such as costs.
(3) Prescriptive analytics use data to identify alternative options and outcomes for project activities.

In addition to the gaps in insights, the use of project data analytics raises questions of moral and ethical responsibility; for example, the direct use of social media communications to represent the voice of stakeholders needs to be carefully managed (Taylor and Pagliari, 2018). Recent work has suggested that the algorithms used in popular text data-analysis techniques can be biased, and the insights generated from them need to be carefully considered (Hovy and Prabhumoye, 2021). Finally, emerging

Table 11.1: Project Data Analytics Use Cases (adapted from Apm.org.uk, 2019).

	Descriptive	**Predictive**	**Prescriptive**
Across projects in a single organization	Stakeholder identity and influence strategies created via social media (Lobo and Abid, 2020)	Social media and litigation risk (Diao et al., 2020)	Causal analytics of social media influence on project outcomes (Nik-Bakht and El-Diraby, 2020)
Across projects in multiple organizations	Dashboards of sentiment of online stakeholder conversations (Tang et al., 2020)	Construction and social media activity (Palmer and Udawatta, 2019)	No extant information to date

demands for reproducibility and open science require that data and the tools that have been used to generate insights are available for inspection and reuse. This approach provides the oversight that supports the ethical use of project data analytics.

11.2 Responsible Research

Responsible research builds on philosophical work in research ethics and anticipatory governance, incorporating a systems perspective that seeks to minimize unanticipated negative outcomes from research (Guston and Sarewitz, 2002). Responsible innovation seeks to

(1) Increase the net societal value created by research by encouraging researchers to address grand challenges.

(2) Improve the process of co-creating research outputs by encouraging active engagement with societal stakeholders before, during, and after research activities.

(3) Reduce the societal risks from research by anticipating potential problems and incorporating impact on social and environmental systems (Fitjar et al., 2019).

These objectives are based on the principles of ethics, equality anticipation, reflexivity, inclusion, and responsiveness and sustainability (Stilgoe et al., 2013), which are similar to those espoused by responsible management (Radoynovska et al., 2020) and responsible project management. These principles may, however, be difficult to enact in project management research. A significant amount of research is conducted by industry that may find inclusion problematic due to intellectual property protection requirements. At the same time, academic researchers may not consider anticipated outcomes due to the need to publish for promotion and other purposes.

Open science is one such approach that attempts to embed responsible research in institutions' approach by providing visibility or transparency to improve oversight (Owen et al., 2019). In addition to scrutiny, open science seeks to improve confidence in research by encouraging reproducibility, defined as performing the same analysis with different researchers and obtaining similar results (Patil et al., 2016). Open research transforms previously closed linear knowledge production systems to encourage networked collaboration. The result can support engagement with the research community and broader society before publishing and greater post-publication engagement and impact. In addition to these benefits, the reuse of data reduces the cost of subsequent research. Research teams are also encouraged to pool their efforts to address grand challenges, such as climate change (McGahan et al., 2021). Open science, however, also depends on the sharing of datasets which can raise challenges, as identified earlier. These can include institutional structures and individual researcher incentives since citations of datasets may be less frequent than those of academic papers and reduce their utility as a metric of research impact in some institutions (Peters et al., 2016).

The temporal nature of project phenomena may be another barrier to open research. However, it is difficult for the domain to make causal claims for research findings without independent confirmation of measures and outcomes. The recent debate about the nature of megaproject cost overruns is illustrative since even though the outcomes (cost higher than budgeted) have been agreed upon, the causes are disputed by researchers (Love et al., 2019).

11.2.1 Responsible Research and Reproduction

A helpful term that may bring clarity to this issue is 'systematic capitalisation on chance', where the career requirement to publish in highly ranked journals encourages researchers to create outputs that show statistically significant findings (Aguinis et al., 2017). While the production may be publishable, the analysis may not be reproducible. Researchers may perform several steps that reduce reproducibility, including biases in design, conceptual underpinnings, research question formation, data collection, analysis, and reporting. These practices can include selective consideration of variables where researchers choose measures based on the outcomes they seek for publication rather than on a conceptual or theoretical basis. This issue is further complicated in project management as the same term may have different meanings in different contexts. For example, project performance has been defined as rankings and ratings

(Szatmari et al., 2021), a seven-measure variable that includes stakeholder satisfaction (Zhu et al., 2021) and a four-variable measure that does not (Hwang et al., 2020). From a reproducibility perspective, it would be difficult to determine if multiple project performance measures were considered before selecting a set of metrics. Further, the popularity of analytical approaches, such as Structural Equation modelling and, more recently, machine learning, provides the opportunity to adjust models to the dataset's characteristics to obtain a publishable fit.

Further, hypotheses and research propositions can also be changed due to the data analysis, resulting in results that fit the data but are not conceptually driven. Finally, many quantitative studies do not report outliers and the management of these cases as part of the analysis. As a result, the veracity of the claims made from these studies is difficult to confirm.

Quantitative researchers in other domains have begun to incorporate approaches to address these issues (Hüffmeier et al., 2016). Table 11.2 provides a description of approaches that quantitative researchers can adopt: (1) exact replications are performed of analyses done by the same researchers ensuring validity by protecting against false positives when findings are unexpected or based on current theory; (2) close replications repeat the same process but using different researchers; (3) constructive replication repeats the previous study with additional elements.

Table 11.2: Approaches for Quantitative Researchers (adapted from Hüffmeier et al., 2016).

	Exact Replication	**Close Replication**	**Constructive Replication**
Authorship	By the same research team	Different research team	The same or different research team
Similarity to the previous study	The same	Same except for researchers	Same conceptual basis, differing conditions
The typical situation for replication	Consecutive study in a paper containing multiple studies, for example, multiple surveys of the same stakeholders	Confirming and evaluating the empirical finding identified in the previous study	Considering empirical findings along with determining the necessary conditions for these findings
Reduces	The potential for false-positive effects in hypothesis testing	(a) False-positive effects (b) Researcher adaptation of model/ hypotheses to fit dataset	(a) False-positive effects (b) Researcher adaptation (c) Impact of external conditions

A significant amount of project management research is qualitative. Due to the open-system nature of projects, qualitative research is valuable in determining the limits to quantitative findings and new concepts to inform theory and practice. The temporal nature of project activities also indicates that both entities and their attributes are constantly evolving; time is a critical factor, and the overall pattern of events determines causality.

While direct replication may not be possible, qualitative research in projects has performed quasi-constructive replications by the site to identify how routines are created and embedded (Davies et al., 2018). Positivist qualitative research, such as case studies can be examined by validity (construct, internal and external). Researchers should seek to be transparent to increase confidence in the research and its findings. At the core, they seek to clarify to research stakeholders why a study was done, why a particular context was selected, how the study was conducted and how the data was analyzed (Pratt, 2008). This domain requires explaining what was done, why it was done, and any changes to the data collection approach done during the research.

11.3 Social Media in Project Management Research

Both quantitative and qualitative approaches have been applied to the relatively new data obtained from social media. These socio-technical systems have generated a significant amount of multimodal data. In addition to audio, video and text, they generate metadata, including personal characteristics, device characteristics, time and location data. The relationships that facilitate social media creation and exchange are traceable via the data and metadata generated by the digital mediums. This data enables the creation of entirely new research questions for project management research. It is possible, however, to examine macro-, meso- and micro-level project network perspectives using data from social media. These platforms facilitate participation by a wide range of participants and provide opportunities to examine project phenomena on a scale that was not possible by using survey or interview-based data collection methods. The advantage of this approach is that researchers can analyze large-scale, real-world networks instead of working with a subset of possible stakeholders. Since it is possible to observe stakeholder behavior in near real-time, social media also enables the examination of activities without delays in interview and survey-based data collection activities. This facilitates the identification of research insights quickly, an outcome of value to both research and practice.

Project management researchers have recognized the value of social media as a research domain and have generally deployed it in three ways.

The first approach examines the use of social media by respondents, and the second utilizes social media metadata to generate research insights, while the third attempts to examine the content of social media postings.

11.3.1 Social Media as a Source of Respondents

These approaches have sought to understand project activities, impacts, and perceptions via participants recruited via social media (Frew, 2019). Several researchers have used surveys to collect data from social media users (Senders et al., 2013). Survey instruments distributed via social media have been used as part of the data collection process to examine stakeholder perceptions of socio-economic barriers to recovery after a disaster. They have also been used to identify specific stakeholders for the study. In this mode, social media was used to provide contact information to support survey targeting and delivery (Jelodar et al., 2016). Survey data have also been combined with social media profile information. For example, the Missouri Department of Transportation has used Facebook advertisements to distribute a survey to residents to evaluate the performance of projects. This research combined quantitative measures with demographic information obtained via Facebook profile (age, gender). Social media has also been used to obtain interview subjects to examine how construction companies engage in social customer relationship management (Swarts et al., 2016).

11.3.2 Direct Analysis of Social Media Text

Beyond obtaining information from social media users, researchers have directly examined the postings on social media. In this paradigm, researchers have used manual content analysis, automated text analysis, and, more recently, images uploaded by users. This approach has also been combined with interviews to examine governmentality in complex construction projects (Ninan et al., 2019).

11.3.3 Automated Text Analysis

Since social media can generate large posts, manual text analysis can only examine a sample of the data posted. Text mining or automated text analysis has been applied to user-generated content posted on social media to overcome these limitations. Text analytics is currently being used as a means of overcoming the limitations mentioned above by using software applications to undertake large-scale analysis of postings, to get an overall sense of topics discussed and their relative popularity. This technique has been used to understand the topics discussed online in the construction industry by using topic modelling of Twitter postings (Tang et al., 2017). A related stream of this work attempts to understand the latent meaning of text via opinion mining—an automated approach that seeks to understand the direction of sentiment (positive or negative) and the nature of interests expressed in text content (Wang et al., 2019).

11.3.4 Metadata

In addition to the content of social media postings, data associated with social media posts, also known as metadata, have been used to generate project-research insights. The geographic metadata associated with tweets has been used to understand patterns of stakeholder interest at a given location, such as the relationships between infrastructure resilience and social equity. Researchers can combine the content and metadata from Tweets with infrastructure damage data and public social equity data. One limitation of social media as a public stakeholder analysis tool is that socially vulnerable populations may be less active on social media and therefore, may not be adequately represented.

11.4 Social Network Analysis and Social Media Research

Project activities are delivered via several activities and interactions between stakeholders, on examining from a relational perspective. In Social Network Analysis (SNA), the distribution and configuration of relationships are used to explain interactions between entities. These entities can take the form of animate and inanimate objects (Scott, 1988). For the former, SNA research examines interactions between individuals, groups, and other entities; in the latter domain, SNA has been used to uncover patterns of interactions between countries, companies, and distribution networks. This contrasts with the dominant variable paradigm in existing social media research that seeks to explain outcomes via associations between entity characteristics.

SNA identifies entities (individuals, groups, countries, cities) as nodes and relationships (information, social, financial), connecting them as edges. This model can be examined mathematically using techniques drawn from matrix algebra or qualitatively to visualize patterns of interactions. Further, network analysis using social media data has been relatively ignored in project management, despite its rapid adoption elsewhere (Williams et al., 2015).

For social network analysis research using social media, Twitter has often been used. While Twitter is not the most heavily trafficked social media website, it has uniquely suitable characteristics for network research. The first is that Twitter follows a real-time streaming format, unlike Facebook, which uses algorithms to determine which content is visible to the user. In this way, researchers can obtain near real-time insights into behavior (Bruns and Stieglitz, 2014).

Further, Twitter posts are public by default unless a user has chosen to limit access to approved followers or blocked offending users purposefully. In addition to user information interactions, individual Twitter posts hold a significant amount of metadata, including geographic and temporal data used to understand node-level characteristics in social networks (HerdaĞdelen et al., 2013).

At the macro level, this research examines how structural properties influence interactions and outcomes in the network (Carrington et al., 2005). Size is simply the number of nodes in the network, while density measures overall network cohesion and compares the number of relationships to the maximum possible number of ties in a network. A higher density level is associated with higher information sharing, collaboration, and support in the network (Granovetter, 1973). Modularity helps describe the degree to which the network can be subdivided into smaller groups or clusters. A high degree of modularity is an indicator that distinct clusters exist in which nodes have a higher level of connection with each other (within the cluster), rather than outside of the cluster (the rest of the network) (Borgatti et al., 2009). Finally, distance examines how nodes that may not be adjacent can connect. A high network distance indicates that exchanges or other interactions need to traverse several nodes to get from one side of the network to another. Further, in these structures, nodes closer to each other may exert more influence than more distant nodes.

At the meso level, distinct subgroups may exist in networks with a medium to high level of modularity. These groups may or may not overlap. If groups overlap, the members may engage in cooperative behavior and information, and other interactions can travel quickly across the entire network. If they do not, conflict can occur in the network, and interactions may be constrained to travel to the originating group only.

Beyond network structures and sub-network configurations, the characteristics of nodes are also examined. Most often, the property of centrality, or the relationship of a node to others, is used to identify the role of a node in the network. Central nodes act as information brokers because interactions will need to pass through them to propagate within the cluster or across the entire network. Centrality has been used to identify important entities in virtual or real stakeholder networks (Williams et al., 2015). It has also been used to analyze spatial flows of individuals (Leung et al., 2012). The characteristics of these nodes have been studied as their motivations may influence discussions in the entire network. For example, if commercial providers manage the key nodes (Twitter accounts), they may attempt to shape the discussions to increase sales or manage complaints to achieve organizational goals.

11.5 Responsible PM Social Media Data Analysis Methodology

The framework presented in Table 11.3 is a proposed approach to conducting Responsible Project Management Research using social media that incorporates the ideas of open research. It adapts the ideas published by previous researchers on building management theory (Shrestha et al., 2021) to present a framework for responsible reproducible social media data collection and analysis in project management. Social media data sources tend to have characteristics that do not suit existing statistical or qualitative approaches as the data does not fit existing distributions and has many features or variables. These datasets can be beyond the capabilities of individuals or teams of researchers to analyze manually. However, for open research, these are not necessarily disadvantages. Due to its inherently digital nature, the process of collecting social media data can meet reproducibility requirements as each step can be documented to enable future researchers to duplicate a given researcher's actions.

Further, depending on the platform, the dataset or metadata could be shared to ensure that researchers seeking to replicate a study have obtained the same data. Secondly, the scale of social media data encourages the application of digital analytical tools in which actions are taken on a dataset; for example, the removal of duplicates can be documented. Further, the more recent development of computational notebooks allows researchers to share code, datasets and visualisations, enhancing transparency and responsibility in research.

Little effort has been dedicated to examining how to conduct this process in project management despite the growing usage of social media data. Further, limited work explicitly examines the building of theory using large amounts of data obtained from computational methods or so-called big data or social media. Theory building has generally focused on organizations and individuals working with small numbers or small amounts of data. This requires a change in not just the methodology of theory building but the analytical methods themselves. An opportunity exists in the use of analytical approaches to support theory building. One of the advantages that this approach provides is identifying patterns in the qualitative sense and allowing other researchers to duplicate the pattern creation process. Social media data collection processes can also be duplicated within the constraints of the platform. In this way, aspects of theorising (data collection, analysis, and pattern identification) are reproducible and explainable. Researchers can ensure that the analysis is valid, but the analysis and the data can be shared with other analysts for review and independent verification. Researchers can then discuss or debate the validity of the inferences, and discussions around the interpretations can also be shared, archived, and left open to scrutiny from a wide variety of stakeholders.

Table 11.3: Protocol for Responsible PM Research Using Social Media Data (adapted from Shrestha et al., 2021).

Stage	Actions	Details	Researcher Judgement Involved	Benefits
Pre-register study	Create and upload study protocol to open data platform	After obtaining ethical approval, the team will create a project folder on osf.io to host the pre-registration document, ensuring public availability for examination by the academic community and funding agency. A pre-registered report is an article in which methods and proposed analyses are made available for commentary prior to research being conducted. The output will be disseminated on a project page created on Osf.io and will support engagement with the academic community as researchers can comment on the approach and adapt or apply it in subsequent work.	Document details in the study protocol	By publishing the pre-registered document, the ethics and execution of the project can be examined by stakeholders.
Obtain sample		Twitter data collection methodology and confirm access from the Twitter platform. Tweets will then be obtained from Twitter via the academic access API, which is a cost-effective way of obtaining a comprehensive dataset.	Selection of search terms and keywords for data collection.	Collection of data within the requirements of the social media platform ensures that other researchers can repeat the study.
Remove all unnecessary identification elements				
Stage 1: Splitting the Separate sample samples for pattern detection and the pattern review	Splitting the sample into subsamples.	Depending on the size of the dataset, randomly subsample data into two parts (e.g., 50/50, 80/20) and label them as sample(s) In (for inductive analysis) and sample(s) PT (for pattern testing).	Choice of initial concepts for testing may be shaped by availability, unsystematic observation, weak theoretical conjectures.	Procedure supports internal replication.

Stage 2: Detection of robust and interpretable patterns	Identifying robust associations within sample 'In'	Quantitative network analysis of retweets (seen as an endorsement of an original message) using the R package 'statnet' will be used to identify the overall online network size (number of accounts participating in the discussion), structure (distribution of interactions), and entities (individuals, organizations and institution accounts) who are participating in these stakeholder conversations.	Selection of algorithms and packages.	
	Identifying interpretable and robust associations within sample In	These overall networks will be further subdivided into topic-focused communities of interest that exchange information more frequently than the rest. Within each cluster, keyword analysis (frequency, co-location, polarity) will be performed using the R package Tidytext to identify the stance of Twitter posters.	Identification of interpretable patterns which connect theoretical concepts to data.	
Stage 3: Theory formulation	Construct a theory that explains patterns or research propositions identified.	Thinking and theorising.	Identify a possible theory that explains patterns. Identify possible hypotheses/research propositions for deductive testing.	
Stage 4: Examine theory and possible research propositions in sample 'PT'	Test implications of theory formulated in Stage 3.	Determine if a theory can explain patterns found in sample 'PT'.	Evaluation of results.	This process provides an opportunity for texting patterns within the study and explains the process for future researchers.

In the first stage, after obtaining ethical approval, the research team will pre-register the study. The research team will create a pre-registered study protocol, a time-stamped record of decisions around study design, methods and analysis, documented before data are collected. In this way, the team can obtain feedback from the academic community and increase the research's credibility. Social media data is multimodal and has qualitative and quantitative elements, making it necessary to adjust the analytical approach as more is known. However, by documenting initial decisions, subsequent changes and rationale can be appended to the record to ensure transparency.

Data collection is then done within the regulations of the platform, for example, Twitter provides access via the academic API and Facebook and Instagram provide access to researchers. Care must be taken to follow the guidelines of these platforms, and the researcher should note the requirements/restrictions at the time of study, should replication be required.

Once data is obtained, the team should split the dataset to support the pattern replication process. Depending on the size of the dataset, the dataset should be randomly subsampled into two parts (e.g., 50/50, 80/20). To support the pattern identification and testing process, the first sample should be labelled 'In' (for inductive analysis or pattern identification) and sample(s) as 'PT' (for pattern testing). Analysis will then be performed on first the pattern identification sample and then the pattern testing sample. If the analysis is performed using a programmatic approach, such as R, this supports reproducibility as R and Python workspaces that contain code and data can be shared directly or via repositories.

This approach will complement the small but emerging body of knowledge to improve PM research outputs (Meredith, 2021). While it contains some guidance that can support PM researchers using social media data to improve their outputs, additional work is needed, particularly in ethics. The protocol design presumes that ethical review by an institution and compliance with the social media platform requirements is sufficient. It does not address the ethics of social media communications analysis. Further, the approach utilizes a text analytical approach based on keywords that do not attempt to summarize perspectives. At the same time, this avoids the challenges of neural network-based analyses that exhibit specific biases. Future research that utilizes more complex algorithms will need to incorporate a statement on the potential impact of the usage of these tools.

11.6 Conclusion

As project management research has begun to embrace an analytical perspective using structured and unstructured data, care must be taken to ensure that the emerging insights are valid (Xu et al., 2020). The use of social media data, in particular, provides researchers with data on stakeholders and enterprises that was not easily available before. Care must be taken, however, as the examination of digital trace data may reveal private information that stakeholders did not wish to disclose. An ethical and responsible research process should encourage the adoption of open and responsible research practices, such as pre-registration where the research design can be examined before execution. Ethical and privacy issues can be identified before the research team attempts data collection or examination of a social media archive. Further, the process of analysis can adapt ideas presented in Table 11.3 to support the development of theoretical insights. While the approach presented utilized Twitter and social network analysis, the framework can be modified to incorporate machine learning approaches. In the case of the latter, a potential practice might be the use of multiple algorithms in the pattern identification stage in order to ensure that insights are valid and not merely an algorithmic artefact. Future research can also examine the validity issues that can be obtained by the use of neural network-based analyses that are inscrutable black boxes which cannot be easily audited.

References

Aguinis, H., Cascio, W.F., and Ramani, R.S. (2017). Science's reproducibility and replicability crisis: International business is not immune. *Journal of International Business Studies*, **48**(6): 653–663.

APM. (2021). *What is Project Data Analytics?* https://www.apm.org.uk/resources/what-is-project-management/what-is-project-data-analytics/.

Borgatti, S.P., Mehra, A., Brass, D.J., and Labianca, G. (2009). Network analysis in the social sciences, *Science*, **323**(5916): 892–895.

Bruns, A., and Stieglitz, S. (2013). Towards more systematic Twitter analysis: metrics for tweeting activities. *International Journal of Social Research Methodology*, 16(2): 91–108.

Carrington, P.J., Scott, J., and Wasserman, S. (eds.). (2005). Models and Methods in Social Network Analysis (Vol. 28). Cambridge University press.

Cosentino, V., Izquierdo, J.L.C., and Cabot, J. (2017). A systematic mapping study of software development with GitHub, *IEEE Access*, **5**: 7173–7192.

Davies, A., Frederiksen, L., Cacciatori, E., and Hartmann, A. (2018). The long and winding road: Routine creation and replication in multi-site organizations, *Research Policy*, **47**(8): 1403–1417.

Diao, C., Liang, R., Sharma, D., and Cui, Q. (2020). Litigation risk detection using Twitter data, *Journal of Legal Affairs and Dispute Resolution in Engineering and Construction*, **12**(1): 04519047.

Fitjar, R.D., Benneworth, P., and Asheim, B.T. (2019). Towards regional responsible research and innovation? Integrating RRI and RIS3 in European innovation policy, *Science and Public Policy*, **46**(5): 772–783.

Frew, S.L. (2019). Harnessing the power of social media for disaster risk reduction and the mitigation planning process. pp. 104–122. *In:* Kar, B., and Cochran, D.M. (eds.). *Risk Communication and Community Resilience*, Routledge.

Guertler, M.R., and Sick, N. (2021). Exploring the enabling effects of project management for SMEs in adopting open Innovation—A framework for partner search and selection in open innovation projects, *International Journal of Project Management*, **39**(2): 102–114.

Guston, D.H., and Sarewitz, D. (2002). Real-time technology assessment, *Technology in Society*, **24**(1-2): 93–109.

HerdaĞdelen, A., Zuo, W., Gard-Murray, A., and Bar-Yam, Y. (2013). An exploration of social identity: The geography and politics of news-sharing communities in twitter, *Complexity*, **19**(2): 10–20.

Hovy, D., and Prabhumoye, S. (2021). Five sources of bias in natural language processing, *Language and Linguistics Compass*, **15**(8): e12432.

Hüffmeier, J., Mazei, J., and Schultze, T. (2016). Reconceptualising replication as a sequence of different studies: A replication typology, *Journal of Experimental Social Psychology*, **66**: 81–92.

Hwang, B.G., Ngo, J., and Her, P.W.Y. (2020). Integrated digital delivery: Implementation status and project performance in the Singapore construction industry, *Journal of Cleaner Production*, **262**: 121396.

Jelodar, M.B., Yiu, T.W., and Wilkinson, S. (2016). Relationship-quality judgment model for construction project procurement: A conjoint measurement, *Journal of Construction Engineering and Management*, **142**(7): 04016012.

Leung, X.Y., Wang, F., Wu, B., Bai, B., Stahura, K.A., and Xie, Z. (2012). A social network analysis of overseas tourist movement patterns in Beijing: The impact of the Olympic Games. *International Journal of Tourism Research*, **14**(5): 469–484.

Liu, Y., and Bakici, T. (2019). Enterprise social media usage: The motives and the moderating role of public social media experience, *Computers in Human Behaviour*, **101**: 163–172.

Lobo, S., and Abid, A.F. (2020). The role of social media in intrastakeholder strategies to influence decision making in a UK infrastructure megaproject: Crossrail 2, *Project Management Journal*, **51**(1): 96–119.

Love, P.E., Ika, L.A., and Ahiaga-Dagbui, D.D. (2019). On de-bunking 'fake news' in a post-truth era: Why does the planning fallacy explanation for cost overruns fall short? *Transportation Research Part A: Policy and Practice*, **126**: 397–408.

McGahan, A.M., Bogers, M.L., Chesbrough, H., and Holgersson, M. (2021). Tackling societal challenges with open innovation, *California Management Review*, **63**(2): 49–61.

Meredith, J.R. (2021). HOLEY MOLEY, the poppycock syndrome, and academic drift, *International Journal of Project Management*, **40**(1): 15–18.

Nik-Bakht, M., and El-Diraby, T.E. (2020). Beyond chatter: Profiling community discussion networks in urban infrastructure projects, *Journal of Infrastructure Systems*, **26**(3): 05020006.

Ninan, J., Clegg, S., and Mahalingam, A. (2019). Branding and governmentality for infrastructure megaprojects: The role of social media, *International Journal of Project Management*, **37**(1): 59–72.

Owen, R., Forsberg, A.M., and Shelley-Egan, C. (2019). *RRI-Practice Policy Recommendations and Roadmaps*, D: 16.2.

Palmer, Stuart, and Nilupa Udawatta. (2019). Characterising 'Green Building' as a topic in Twitter, *Construction Innovation*, **19**(4): 513–530.

Patil, P., Peng, R.D., and Leek, J.T. (2016). A statistical definition for reproducibility and replicability, *BioRxiv*, 066803.

Peters, I., Kraker, P., Lex, E., Gumpenberger, C., and Gorraiz, J. (2016). Research data explored: An extended analysis of citations and altmetrics, *Scientometrics*, **107**(2): 723–744.

Pratt, M.G. (2008). Fitting oval pegs into round holes: Tensions in evaluating and publishing qualitative research in top-tier North American journals, *Organizational Research Methods*, **11**(3): 481–509.

Radoynovska, N., Ocasio, W., and Laasch, O. (2020). The emerging logic of responsible management: Institutional pluralism, leadership, and strategising. pp. 420–437. *In:* O. Laasch, R. Suddaby, R.E. Freeman, and D. Jamali (eds.). *Research Handbook of Responsible Management*, Edward Elgar Publishing.

Ruijer, E.H., and Martinius, E. (2017). Researching the democratic impact of open government data: A systematic literature review, *Information Polity*, **22**(4): 233–250.

Scott, J. (1988). Social network analysis, *Sociology*, **22**(1): 109–127.

Seddon, J.J., and Currie, W.L. (2017). A model for unpacking big data analytics in high-frequency trading, *Journal of Business Research*, **70**: 300–307.

Senders, A., Govers, R., and Neuts, B. (2013). Social media affecting tour operators' customer loyalty, *Journal of Travel & Tourism Marketing*, **30**(1-2): 41–57.

Shrestha, Y.R., He, V.F., Puranam, P., and von Krogh, G. (2021). Algorithm supported induction for building theory: How can we use prediction models to theorize? *Organization Science*, **32**(3): 856–880.

Stilgoe, J., Owen, R., and Macnaghten, P. (2013). Developing a framework for responsible innovation, *Research Policy*, **42**(9): 1568–1580.

Swarts, K.M., Lehman, K., and Lewis, G.K. (2016). The use of social customer relationship management by building contractors: Evidence from Tasmania, *Construction Management and Economics*, **34**(4-5): 302–316.

Szatmari, B., Deichmann, D., van den Ende, J., and King, B.G. (2021). Great successes and great failures: The impact of project leader status on project performance and performance extremeness, *Journal of Management Studies*, **58**(5): 1267–1293.

Tang, L., Zhang, Y., Dai, F., Yoon, Y., Song, Y., and Sharma, R.S. (2017). Social media data analytics for the US construction industry: Preliminary study on Twitter, *Journal of Management in Engineering*, **33**(6): 04017038.

Tang, L., Griffith, L., Stevens, M., and Hardie, M. (2020). Social media analytics in the construction industry comparison study between China and the United States, *Engineering, Construction and Architectural Management*, **27**(8): 1877–1889.

Taylor, J., and Pagliari, C. (2018). Mining social media data: how are research sponsors and researchers addressing the ethical challenges? *Research Ethics*, **14**(2): 1–39.

Wang, Y., Li, H., and Wu, Z. (2019). Attitude of the Chinese public toward off-site construction: A text mining study, *Journal of Cleaner Production*, **238**: 117926.

Williams, N.L., Ferdinand, N., and Pasian, B. (2015). Online stakeholder interactions in the early stage of a megaproject, *Project Management Journal*, **46**(6): 92–110.

Xu, H., Zhang, N., and Zhou, L. (2020). Validity concerns in research using organic data, *Journal of Management*, **46**(7): 1257–1274.

Zhou, Z., Zhou, X., and Qian, L. (2021). Online public opinion analysis on infrastructure megaprojects: Toward an analytical framework, *Journal of Management in Engineering*, **37**(1): 04020105.

Zhu, F., Wang, X., Wang, L., and Yu, M. (2021). Project manager's emotional intelligence and project performance: The mediating role of project commitment, *International Journal of Project Management*, **39**(7): 788–798.

12

An Update on Social Media in Project Management

Applications, Developments and New Perspectives

Gilbert Silvius

12.1 Introduction

When reaching out to the project management community for experiences and studies on the use of social media in project management, for the book *Strategic Integration of Social Media into Project Management Practice* (Silvius, 2016a), the response was surprisingly low. Social media had already earned its position in our private and professional lives, and therefore, also in project teams, but a more documented study on the applications, effects, and challenges was hard to find. A few years before that, Elizabeth Harrin had published a first book on the topic (Harrin, 2010), but more serious academic work was still lacking. However, that was in 2015–2016.

Today, more than five years later, the use of social media in our lives has progressed. The novelty and hype have passed and using social media is an integrated part of our personal and professional communication. However, as the use of social media has grown, we have also experienced the downside of it. Along with the benefits of easy connectivity and increased information, social media also provides politically motivated

HU University of Applied Sciences Utrecht, The Netherlands.

disinformation, aimed at undermining or disrupting societal balances. Social media platforms have become both successful and problematic (Aral, 2020). The (lack of) responsibility, that social media platforms take for the messages their users share, is frequently criticized. However, social media users themselves logically also bear a responsibility for the information they share and they take from social media. That means that individuals should make more conscious and deliberate decisions with regards to their use of social media, thereby weighing functionality and value, with effort and risks.

The use of social media in projects implies even greater care and consideration to be given to the use of social media in communication in and by the project. Other than in 2016, it is no longer the question whether a project should use social media, but how it should use it. The present book is therefore called for and its contributors present a rich review of the current state of social media use in project management.

In this reflection, the central question is how the applications of social media in project management have evolved in the past years? What is the status of the use of social media in the earlier identified application areas and which new developments can be observed?

The following sections will first introduce three identified application areas of social media in project management, after which this structure will be used to discuss the contents and new perspectives of the present book.

12.2 The Application of Social Media in Project Management

In a reflection on the earlier mentioned book of 2016, we recognized three areas of application of social media in project management (Fig. 12.1):

1. Social media in team communication and collaboration.
2. Social media in external stakeholder engagement.
3. Social media in project manager's professional development.

The first application of social media in project management, that of supporting team communication and collaboration, was already well established. Team collaboration suites, such as Project place, Basecamp and Redmine, provided many of the functionalities of social media that Kietzmann et al. (2011) identified in their honeycomb model. These platforms typically provide good functionality for several modes of communication, knowledge and information sharing, document management, and coordination of tasks. The more privately used social media platforms, such as Facebook, Yammer, and LinkedIn provided

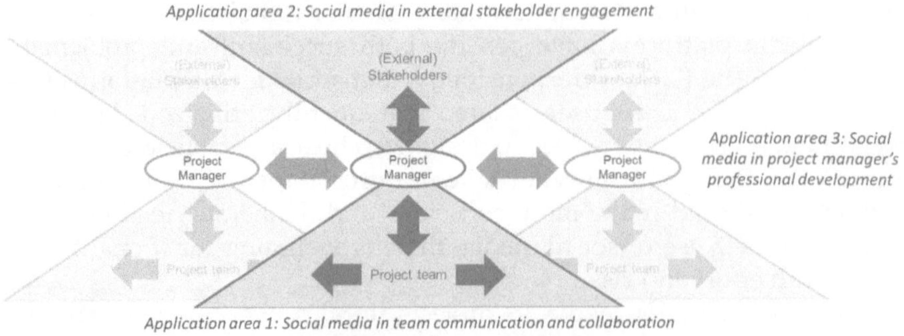

Fig. 12.1: The applications of social media in project management (Silvius, 2016b). (Alt Text: The three application areas of project management are shown with project manager, project team, and external stakeholders)

complementary functionality that allowed their users to build a digital identity and reputation by sharing content and building relationships, which also was expected to stimulate the quality of teamwork in the context of project teams.

The second application of social media in project management is that of external stakeholder engagement. Social media was seen to provide unprecedented opportunities for two-way communication with stakeholders. The traditional approach to stakeholder management, which evolves around 'sending' information to stakeholders in order to hope to win their support, can, with social media, evolve into stakeholder engagement, in which stakeholders are consulted and invited to co-create the project. Stakeholder engagement is an approach to stakeholder relations that build upon an open interaction with stakeholders in order to create a 'better' project for all involved. However, despite this opportunity, it still had to be concluded that the application of social media for the engagement of stakeholders was 'lagging behind' (Silvius, 2016c). Project managers were reluctant to explore the use of social media beyond the project team, with barriers seen in loss of privacy, blurring of lines between professional and personal online presence, and concerns about data security or compliancy with the organization's policies.

A third application that we identified was that of the use of social media in the professional development of project managers and project management. Project teams are increasingly international, intercultural, geographically dispersed, and self-organizing, which provide project managers with the task to facilitate team development, increase collaborative capabilities, and maximize team productivity. Social media was expected to be able to support project managers in their responsibility, and make project management truly social. 'Social project management'

(Silvius, 2016c) recognized that project teams and project managers are part of a broader project community, which is both formal and informal, and that engaging with that community is the key to getting things done (Triolog, 2012). In social project management, the core business processes of the project and project's management are brought online, making the project process visible to everyone, both inside and outside the team. This transparency into the events of the project is expected to enable smarter, quicker, and more efficient collaboration with regards to the execution and management of the project. Social project management allows for anyone who is interested, to engage socially with the team to assist in accomplishing the project and organization's goals (Triolog, 2012).

12.3 Development of the Application Areas

A review of the current state of social media use in project management is presented in the contents of the present book.

The application of social media in team communication and collaboration is still a strong topic, with the chapters of Chaves et al. (Chapter 3) and Jia et al. (Chapter 4) mainly centered around this application. Both chapters highlight the potential role of social media for knowledge management in the project team. And as earlier publications on the use of social media in project teams mainly focused on a task-oriented use of social media in operational communication, coordination, and collaboration, the focus on knowledge management provides a new perspective. Reflecting on this, it might be observed that the application of social media in project teams is developing from being task-oriented to being learning-oriented. Thompson (Chapter 2) also touches upon this learning-orientation when she points out the role of social media in socialization and internalization processes within the project team.

The second application area of social media in project management is that of stakeholder engagement and is covered in the present book in the chapters by Lehtinen and Aaltonen (Chapter 9), Lobo (Chapter 6), Ninan (Chapter 7) and Zhou et al. (Chapter 8). When reflecting on these chapters, it is striking that the empirical studies reported in these chapters are all large infrastructure projects that have a high visibility to the general public. Apparently, the engagement opportunities that social media offer are a good fit with the communication and engagement strategies of these large projects. Given also the presence of other large infrastructure projects, the 2016 conclusion that the application of social media for the engagement of stakeholders is 'lagging behind', can be dismissed. For large infrastructure projects, the use of social media seems to be an integrated part of their communication and stakeholder engagement strategy.

Despite this development, the focus of the chapters on large infrastructure raises the question whether the use of social media for stakeholder engagement is limited to these large and public-facing projects? This question remains unanswered by the chapters in this book. Conceptually there would not be a limitation for the use of social media for stakeholder engagement also in other types of projects. Also, in organizational change or innovation projects, where there is no need to communicate with the general public, social media can be used to engage with larger groups of stakeholders, for example, employees of the project-performing organization. For confidentiality reasons, the engagement with stakeholders within the organization would logically not be done over open social media platforms. However, the social media landscape of today also provides solutions for this, for example, Yammer.

Next to the empirical studies on the application of social media in project stakeholder engagement included in the book, Lehtinen and Aaltonen (Chapter 9) also provide a noticeable theoretical contribution. Their engagement approach/communication mode matrix provides a practical framework for the classification of project organizations based on their use of social media in stakeholder engagement. With this framework, this application of social media in project management can be further studied and thereby developed.

The third identified application area of social media in project management—the use of social media in the professional development of project managers and project management—is addressed by Harrin (Chapter 5). Her report on the experiments with online communities of project managers provides guidance for further experimentation. Interacting with colleagues is an essential element in professional development. Interacting with other project managers is therefore essential for the professional development of individual project managers. This insight should be sufficient motivation for project managers to actively get engaged in interacting with professional colleagues, both internal and external to their employer. However, engaging with colleagues also requires a certain level of trust and respect. Large, open, and anonymous social media platforms, such as Linkedin, may therefore not be the most effective platforms to support professional development, despite their support of content-related group discussions. Smaller, social media-enabled, intimate communities of practice, as reported by Harrin, may be the way to go, but need to be explored further.

12.4 New Perspectives

A new perspective on the use of social media in project management is provided in the chapters of Williams et al. (Chapter 11) and Zhou et al.

(Chapter 8). These two chapters highlight the opportunity to analyze and mine communication about a project on social media platforms. As 'project data analytics' in itself is already an emerging practice, the use of social media by the project and in the project provides an elaborated set of data which can be explored with data mining and data analysis techniques. As an illustration of this practice, the chapter by Zhou et al. describes a case study on how social media data mining was used to analyze the public opinion in a large infrastructure project. The chapter by Williams et al. suggests that data analysis can even be used to predict future project states or events and prescribe actions. This chapter also discusses the moral and ethical aspects of project data analytics.

12.5 Conclusion

The central question in this reflection was how the applications of social media in project management have evolved in the past years? With regards to the use of project management in team communication and collaboration, we reflected that the contributions in this book represent how this area of application is developing from being task-oriented to being learning-oriented.

With regards to the use of social media in stakeholder engagement, we observed that the empirical studies reported in the book describe social media use in large, public facing infrastructure projects. The use of social media in the stakeholder engagement of smaller and other types of projects appears to be still largely unexplored. The engagement approach/communication mode matrix, developed by Lehtinen and Aaltonen, provides a practical framework for further studies on this application area.

With regards to the use of social media in the professional development of project managers and project management, the contribution of Harrin provides guidance for further studies and experiments with smaller, social media-enabled, intimate communities of practice.

A new perspective that emerged from the book is that of project data analytics on social media communication, which can provide descriptive insights into stakeholder's opinions, and potentially also predictive or prescriptive insights.

References

Aral, S. (2020). *The Hype Machine: How Social Media Disrupts Our Elections, Our Economy, and Our Health and How We Must Adapt*, Penguin Random House: New York.

Harrin, E. (2010). *Social Media for Project Managers*, Newtown Square: PMI.

Kietzmann, J.H., Hermkens, K., McCarthy, I.P., and Silvestre, B.S. (2011). Social Media? Get serious! Understanding the functional building blocks of social media, *Business Horizons*, **54**: 241–251.

Silvius, A.J.G. (2016a). *Strategic Integration of Social Media into Project Management Practice*, IGI Global.

Silvius, A.J.G. (2016b). Social Project Management? Keynote presentation on the *Fifth International Scientific Conference on Project Management in the Baltic Countries*, Riga.

Silvius, A.J.G. (2016c). Social project management, *PM World Journal*, V (VII).

Triolog. (2012). *Social Project Management: Engaging the Social Network to Deliver Project Success*, Triolog Group.

Index

Contributors Bio

Johan Ninan (Dr.) is an Assistant Professor at Delft University of Technology (TU Delft), Netherlands. Previously, he was a research associate at The Bartlett School of Sustainable Construction at University College, London (UCL), UK. His research focuses on megaprojects, stakeholder engagement, innovation, and project organizing with a particular emphasis on the role of digital media. He had hands-on experience in the planning and construction of multiple infrastructure megaprojects prior to delving into researching the project management complexities surrounding such projects in Australia, India and the UK. He was awarded the 2020 IPMA Global Young Researcher Award and the 2020 APM Paper of the Year Award for his research on social media in projects.

Karen Thompson, PhD, is a Senior Academic at Bournemouth University and co-founder of the Responsible Project Management movement. She is an innovative educator and researcher working at the intersection of project management and sustainable development. Her early career was in information technology and systems development. Karen's current research interests include the development of new competencies for responsible management, stakeholder engagement in projects, and partnership development for achieving Carbon Net Zero. She is a Fellow of the Association for Project Management, Fellow of Advance HE, member of the Project Management Institute and expert at the Centre for Project Management at the University of Jaffna, Sri Lanka.

Marcirio Silveira Chaves is an Associate Professor in the Business School at Pontifical Catholic University of Rio Grande do Sul (PUCRS), Brazil where he leads the Managing Projects and Information Technology—MProTech research group. He holds a Ph.D. in Informatics from the University of Lisbon, Portugal. His research findings have been published in over 100 peer-reviewed papers, with more than 1,400 citations in reputed outlets and conferences. His current h-index is 20. He also has experience as a coordinator in national and international research projects funded by Brazilian and European official agencies.

Giuliana Veronese is currently the Digital Solutions Center VP at Stefanini Europe. She is responsible for co-creating strategies and roadmaps for digital transformation. Her responsibilities also include keeping the products and services portfolio updated with new market trends, integrating various Stefanini group solutions, and fostering an agile culture with a growth mindset. Giuliana has a solid tech background; she holds a Ph.D in Computer Engineering from the University of Lisbon. With over 20 years of experience, she has built an international career working in several countries: Portugal, Mexico, Peru, Colombia, Argentina, the United States and Romania.

Roberto Antonio Winter holds a Master's degree in Project Management from Universidade Nove de Julho (UNINOVE) and an MBA in Strategic Management in IT from FGV. He has been an IT professor working for the last three years in an NGO to train people with social vulnerability. Roberto is also a teaching coordinator at the Paula Souza Technical School. He has been an IT professional for 30 years, with experience in various industries, with the last 17 years in insurance companies, in an executive and technical leadership position. He is a professional focused on teamwork and strengthening the 'Soft Skill' profile.

Jianyao Jia is a Ph.D candidate for management science and engineering at the Tongji University, Shanghai, China. His interests include application of mobile ICT to improve project performance, technology, and organizing. His research has been published in peer-reviewed journals, such as the *Automation in Construction, Journal of Construction Engineering and Management, Journal of Management in Engineering*.

Guofeng Ma received the Ph.D degree in management from the Shanghai Jiaotong University, Shanghai, China, in 2004. He is a full Professor for Management Science and Engineering at the Tongji University. He is interested in application and management of digital technologies. In his work, he focuses on how to leverage the benefits of digital technologies as well as mitigate potential risks induced by digital technologies. His work has appeared on *International Journal of Project Management, International Journal of Quality and Reliability Management*, et al.

Shan Jiang is a Ph.D candidate for Management Science and Engineering at the Tongji University, Shanghai, China. He is interested in technology management in projects. His work has appeared on *Engineering Construction and Architectural Management*.

Elizabeth Harrin holds degrees from the University of York and Roehampton University and has published several books on project management. She is a Fellow of the Association for Project Management. As a practitioner, Elizabeth has over 20 years of diverse experience in business change and technology projects predominantly in the healthcare and financial services industries. She's Director of Otobos Consultants Ltd, a consultancy that supports project-led businesses with training, mentoring, and content design where she specializes in helping teams manage multiple projects, engage stakeholders, and still leave the office on time.

Sunila Lobo (Dr.) is an academic with international experience in areas including leadership, management, and marketing across the construction, engineering, and creative industries and more recently, in the third sector. She has an abiding interest in the digital side of business, including how public stakeholders attempt to influence decision-making in megaprojects through use of social media, building a collective identity for their cause. She is a Liveryman of the Worshipful Company of Marketors of the City of London, providing consultancy for outreach clients, thought leadership on current topics, while also conducting research to contribute to social justice. She has recently been appointed a Research Affiliate with the University of Sydney's John Grill Institute for Project Leadership.

Yongjian Ke is an Associate Professor in Project Management in the School of Built Environment at the University of Technology Sydney in Australia. His research spans a wide range of problems in delivering infrastructure projects, focusing on risk allocation and management, contracting behaviors, social acceptance, and social sustainability. He is particularly interested in investigating public-private partnership and its implementation in the infrastructure sector. His work has recently focused on using social media in project management and applying the concept of 'social license to operate' in infrastructure projects to improve community engagement. As of December 2021, Dr. Ke has published 123 papers, which have generated 5,059 citations with an h-index of 31 and an i10-index of 50.

Shankar Sankaran is the Professor of Organizational Project Management at the School of the Built Environment. Shankar's research covers organizational project management, megaprojects, systems thinking, project leadership, and action research. His current research projects are: Practice-based study of project portfolio management and strategic alignment in construction and social media use in project management—multiple case studies of transport projects. He is on the editorial board of the *International Journal of Project Management*, *The International Journal of Managing Projects in Business and Systems* and edits *Project Management Research and Practice*. He is a Visiting Professor in the Erasmus Mundus Programme at the Umea Business School in Sweden; Associate of the Chair of Project Management at the University of Quebec at Montreal (UQAM), Canada; past President of the International Society for the

Systems Sciences; Chair of the Global Accreditation Centre of the Project Management Institute's board (from 2021).

Sandeep Mathur is a Director—Active Transport at Transport NSW, where he manages portfolio, data & analytics functions. He is also Fellow of Australian Institute of Company Directors; Fellow of Australian Computer Society and an Industry Fellow at University of Technology, Sydney. He has over 20 years of project, programme, and portfolio management experience. He has been on several for-profit and not-for-profit Boards in past 20 years. He is a guest lecturer at UTS Faculty of Design, Architecture & Building and also doing Doctoral Research on the topic 'Managing Data Science Initiatives as Exploratory Projects: A New Approach to Programme Management'.

Lauri Vuorinen, D.Sc. (Tech.), works as Post-doctoral Researcher at Tampere University, Finland. Vuorinen's field of research is project business. He has recently focused on topics like stakeholder relations, stakeholder engagement, and value creation in temporary organizations. Vuorinen's research has appeared in publication outlets, such as *International Journal of Project Management* and *International Journal of Managing Projects in Business*.

Ganesh Devkar is currently working as Associate Professor at the Faculty of Technology, CEPT University, Ahmedabad. He completed his Ph.D from Indian Institute of Technology, Madras (IIT Madras), India, and his doctoral thesis focuses on identifying the competencies necessary for effective implementing Public Private Partnerships (PPPs) in infrastructure service delivery and developing a framework that allows urban local bodies to assess their capability to successfully deliver projects through PPPs. His research interests are in the areas of lean construction, PPPs, and urban infrastructure development.

Zhipeng Zhou (Dr.) is an Associate Professor from Department of Management Science and Engineering at Nanjing University of Aeronautics and Astronautics (NUAA). He earned a Bachelor's degree in survey engineering and a Master's degree in Construction Engineering and Management from Southeast University, and a joint Ph.D with a focus on construction safety risk from Southeast University—Georgia Institute of Technology. Before joining NUAA, he worked as a research assistant in the University of Hong Kong and as a Research Fellow in National University of Singapore. His principle research interests are megaproject management, construction safety management, and accident analysis.

Xingnan Zhou (Ms) is a Master student at Nanjing University of Aeronautics and Astronautics (NUAA), majoring in management science and engineering. She earned a Bachelor's degree in Information

Management and Information System from Jiangsu University of Science and Technology. She obtained the 2019–2020 Merit Postgraduate and the 2021 National Scholarship. Her principle research interests are online public opinion analysis and megaproject management.

Lingfei Qian (Dr.) is an Associate Professor at Department of Management Science and Engineering at Nanjing University of Aeronautics and Astronautics (NUAA). She earned a Bachelor's degree in automation engineering and a Master's degree in computer science and technology from Beijing Insitute of Technology. She earned a Ph.D in Information Management from Nanjing University. Her principle research interests are knowledge organization and service, public opinion analysis, and manufacturing intelligence.

Haonan Qi (Ms) is a Ph.D student at Nanjing University of Aeronautics and Astronautics (NUAA), majoring in Management Science and Engineering. She earned a Bachelor's degree in Chemistry from Hebei Normal University and a Master's degree in Chemical Engineering from Southeast University. Her principle research interests are at text-based safety risk analysis, construction project management, social responsibility, machine learning, construction safety performance evaluation.

Jere Lehtinen (Dr. Tech.) is a post-doctoral researcher at both University of Oulu, Industrial Engineering and Management Research Unit (Project Business Research Team), Finland, and Aalto University, Department of Industrial Engineering and Management (Project Business Research Group), Finland. His research focuses on stakeholder engagement, collaboration, and project governance in complex and inter-organizational projects. He also studies project value creation and networks, and the roles of novel technologies, simulations, and games in project management. His work has been published in *International Journal of Operations and Production Management*, *Industrial Marketing Management*, and *International Journal of Project Management*.

Kirsi Aaltonen (Dr. Tech) is Associate Professor of Project Management at University of Oulu, Finland, where she heads Project Business Research Team in the Industrial Engineering and Management Research Unit. She has published widely on management of complex projects and project stakeholder management and engagement in *International Journal of Operations and Production Management*, *Industrial Marketing Management*, *Scandinavian Journal of Management*, *International Journal of Project Management*, *Project Management Journal*, and *International Journal of Managing Projects in Business*. Her current research interests include stakeholder management in complex projects, governance of inter-organizational projects, game-based learning methods, and institutional change in project-based industries.

Nigel L Williams (Dr.), PMP is a Reader in Project Management and Research Lead within the Organizations and Systems Management Subject Group at the Faculty of Business and Law at University of Portsmouth, United Kingdom and co-founder of Responsible Project Management (www.ResponsiblePM.com). His Ph.D is in Engineering from the University of Cambridge, UK, and before joining academia, he worked in the energy industry and as a business consultant for enterprises in the Caribbean. He is Co-editor of the *Handbook of Responsible Project Management* and is the Deputy Editor of *Project Management Research and Practice Journal.*

Nicole Ferdinand is Programme Lead, post-graduate programmes at the Oxford School of Hospitality Management, Oxford Brookes University. She regularly publishes in the areas of tourism, culture, events and project management. She holds a Ph.D in Culture, Media and Creative Industries from King's College, London, as well as an MSc in Marketing and BA in English from the University of the West Indies, St. Augustine.

Beverly Pasian (Dr.) has a 25+ year career in the public sector and higher education based on the pillars of online education, project management and now, smart cities. To maximize her expertise in these areas, her education includes two Master's degrees (in education and business), a Doctorate in Project Management (2011) and a Doctorate in Business Administration (2023). To fulfil the three core jobs of an academic—teaching, research and service—Beverly has worked within a robust international network over the last 10 years. As the designer, developer or lecturer of more than 50 courses, she has taught in a dozen countries, delivered more than 100 papers and presentations at global events, and served in leadership of multiple professional associations. 2022 will see the publication of multiple books. Understanding the role of projects in smart city quality of life has become her research priority with an international case study launched in 2022, starting with the leading city of Bilbao. A new senior research role at FH-Dortmund, Germany, supports these efforts.

Arash Daemi is a Lecturer in Asset and Maintenance Management at Central Queensland University, Australia, where he is also a Ph.D candidate. He is passionate about innovative teaching tools which deliver a unique learning experience to project management students. He has developed project management courses for certified education providers and received student voice commendation and Educator of the Year Award for his professional interactions with project management students. Arash is a PRINCE2 certified project management practitioner. He has managed manufacturing, supply chain, and organizational change projects for more than a decade. He is a member of the Project Management Institute.

Ritesh Chugh is at the School of Engineering and Technology in Central Queensland University, Australia. As an information systems socio-technological expert, his research focuses on the social role of emerging information systems and their influence on humans and organizations in real-world settings. He takes an inter-disciplinary approach in research under the overarching theme of information systems to include socio-technical aspects of social media, knowledge management, project management, educational systems, technology-enhanced learning. He has received several teaching awards that recognize his teaching excellence, commitment to improved student outcomes, and engagement in reflective learning and teaching activities.

Muralitheran V Kanagarajoo is the Head of the Asset and Maintenance Management course at Central Queensland University, Australia. He has a commercial background spanning 20 years that includes business management and business consultation in organizations globally. His research interests lie in the areas of strategic business management, IT planning, and the use of social media in project management. Murali is a certified Lean Six Sigma Consultant and accredited as PMI's project management professional.

Gilbert Silvius is an authentic thinker, an experienced lecturer, and an innovative author. He currently holds academic positions at HU University of Applied Sciences, Utrecht, Wittenborg University of Applied Sciences, LOI University of Applied Sciences and University of Johannesburg. Gilbert is engaged by several other universities in programmes on project management, business administration, and information management. Gilbert is a Fellow of the UK Higher Education Academy, and Honorary Fellow of Turku University of Applied Sciences and the International Information Management Association. Gilbert authored or edited 13 professional and academic books and over 100 academic papers. Gilbert is recipient of the IPMA Outstanding Contribution Research Award (2020) and the GPM Sustainability Award (2013). Next to his academic work, Gilbert advizes organizations as part of the enable2change network of experts. Gilbert holds a Ph.D degree in Information Sciences from Utrecht University and Master's degrees in Economics and Business Administration. He is also a certified project manager, scrum master, and product owner.